\. S.
ìr
ì)

, the
f the
verse.

Edwin Hubble was one of the outstanding astronomers of the 20th century, who in a series of observations made in the 1920s discovered the expansion of the universe. Hubble opened the world of galaxies for science when he showed that spiral nebulae beyond the Milky Way are galaxies extending to the limits of the universe, and participating in a general expansion of the cosmos. The exploding universe of Hubble, now termed the Big Bang, determined the origin of the elements, of galaxies and of the stars.

This biography is the first complete account of the scientific life and work of Hubble. Significant family documents relating to Hubble are published here for the first time. The book gives a detailed description of the activities of this famous scientist, whose discoveries firmly established the United States as the leading nation in observational astronomy. The story is enriched by Alan Sandage, who worked with Hubble at the Mount Wilson and Mount Palomar Observatories.

The second part of this book describes the fundamental discoveries on the nature of the universe made subsequently, and thus sets his achievements in context. The result is a book that is a real classic of science, setting out the thrilling story of the exploding universe.

The authors are two prominent astronomers who have built on Hubble's work. Alexander S. Sharov is a researcher at the Sternberg Astronomical Institute, Moscow, who has worked on the nearby galaxies first studied in detail by Hubble. The cosmologist Igor D. Novikov is now at the prestigious Nordita research institute in Denmark, concentrating on black hole physics and the expanding universe.

Edwin Hubble, the Discoverer of the Big Bang Universe

Frontispiece. Edwin Powell Hubble 1889–1953.

EDWIN HUBBLE,
THE DISCOVERER
OF THE BIG BANG
UNIVERSE

ALEXANDER S. SHAROV
P. K. Sternberg Astronomical Institute, Moscow

and

IGOR D. NOVIKOV
Nordita, Copenhagen

TRANSLATED BY VITALY KISIN

CAMBRIDGE
UNIVERSITY PRESS

Published by the Press Syndicate of the University of Cambridge
The Pitt Building, Trumpington Street, Cambridge CB2 1RP
40 West 20th Street, New York, NY 10011–4211, USA
10 Stamford Road, Oakleigh, Melbourne 3166, Australia

Originally published in Russian as *Chelovek, otkryvshiĭ vzryv
Vslennoĭ* by Nauka (Science) Publishing House, Moscow, 1989
and © Nauka Publishing House 1989

First published in English by Cambridge University Press
English edition © Cambridge University Press 1993

Printed in Great Britain at the University Press, Cambridge

A catalogue record for this book is available from the British Library

Library of Congress cataloguing in publication data

Sharov. A. S. (Aleksandr Sergeevich)
   [Chelovek, otkryvshiĭ vzryv Vselennoĭ . English]
   Edwin Hubble, the discoverer of the Big Bang Universe /
Alexander S. Sharov and Igor D. Novikov: translated by Vitaly Kisin.
      p.   cm.
   Includes bibliographical references and indexes.
   ISBN 0–521–41617–5
   1. Hubble, Edwin Powell, 1889–1953. 2. Astronomy–History–20th century.   3.
Cosmology–History–20th century. 4. Astronomers–United States–Biography.   I. Novikov.
I. D. (Igor Dmitrievich) II. Title.
QB36.H83S5313   1993
520′ .92–dc20
[B]   92–24683   CIP

ISBN 0 521 41617 5 hardback

CONTENTS

Preface to the English edition	ix
Preface to the Russian edition	xi
Part 1: Life and work	**1**
Choosing the way	1
On the threshold of great achievements	14
The island universe	26
The red-shift. Predecessors	47
Hubble's Law	56
Recognition. Facets of his personality	74
Civil duty	85
Hopes crushed	102
Part 2: Hubble's work continued	**123**
Distances to galaxies and the Hubble constant	123
Current and future research projects	133
The discovery of the hot universe	140
Explosion	148
Chronology of Edwin Hubble's life and work	168
Bibliography of Hubble's publications	170
References	176
Index of names	179
Index of celestial objects	183
Subject index	184

Preface to the English edition

Years and decades have passed since Hubble's classical work. There is no question that he was the greatest observational astronomer since Copernicus. The three enormously important things he did were: he discovered galaxies, he showed that they were characteristic of the large-scale structure of the universe, and then he found the expansion. Any one of those is monumental and would secure his place in history.

Alan Sandage
From the interview for this book, given to Kip Thorne

The astronomical community did not overlook the centennial of the most outstanding astronomer of the 20th century, Edwin Hubble. On 21–23 June 1989, a symposium on the evolution of the universe of galaxies was convened in Berkeley, CA, dedicated to Hubble's memory. Three of the papers read at the symposium were devoted to the life and work of the great scientist. In Moscow, one of the authors (A. Sharov) presented a talk at about this time on Hubble's life and work, at the P. K. Sternberg Astronomical Institute in Moscow. In all likelihood, people similarly remembered Edwin Hubble in 1989 at other institutes and observatories. Several periodicals published articles about him.

Among the things said and written about Hubble on the occasion of the centenary, the most interesting was the talk delivered at the Berkeley symposium by D. E. Osterbrock, R. S. Brasher and J. A. Gwinn. These authors had worked through a staggering amount of archive material (81 references) and went through Hubble's life thoroughly, from his date of birth to the middle 1920s. They discovered information which had been completely unknown to everyone, including the authors of the present book. The text of this talk appeared in *The Evolution of the Universe of Galaxies – Edwin Hubble Centennial Symposium* (Berkeley, CA, 21–3 June 1989, edited by R. G. Kron, Conference Series Volume **10** (1990) San Francisco: Astronomical Society of the Pacific, pp. 2–14).

Nevertheless, there is still no detailed biography of Edwin Hubble, which is so needed and which would describe his entire life and his outstanding achievement. This is why we dare to offer to the general public our book, written and published in Moscow on the occasion of Hubble's centennial and now translated into English. Changes and additions to the text of the first part of the book, describing the life and work of Edwin Hubble, are quite small as compared with the Russian edition. We decided not to introduce into our text the information presented in the article of D. E. Osterbrock, R. S. Brasher and J. A. Gwinn. We simply invite the reader to read this excellent article too. Reading both texts would help one to get a better picture of Hubble's personality, his life and his work. The second part of the book, dealing with the expansion and continuation of the work to which Edwin Hubble devoted most of his life, has been enlarged to a considerably greater extent.

The second half of the 20th century is characterised by a constant intensification of the pace of life. It is not surprising that further progress in astronomy and space research in the three years since the publication of the Russian edition has brought us a huge amount of new knowledge. We have made an attempt to incorporate these new developments in the English edition.

A. Sharov
I. Novikov

Preface to the Russian edition

A hundred and fifty years ago, half a century before the protagonist of this book was born, the great Russian poet Mikhail Lermontov wrote that 'A preface is the first but also the last and the poorest thing in any book – it either explains the purpose of the book or serves as its justification and an answer to critics'. Many things have changed in this world since then but writing is still the same kind of work. Authors still cling to writing prefaces. The present authors are no exception. Lermontov complained that readers were bored with prefaces. In our dynamic age, this must be especially so. Still, the reader is advised not to hurry on and to read these pages before leafing through the main text.

The purpose of the book is clear. It is to describe the life and work of the American astronomer Edwin Powell Hubble (1889–1953) whose centenary was celebrated quite recently. The history of astronomy concerns many illustrious men who richly deserve the grateful memory of mankind. Edwin Hubble was outstanding even among this group.

Astronomers have a fine tradition of naming large telescopes after famous scientists – the 4-metre Newton telescope, the 5-metre Hale telescope, the G. A. Shain reflector in the Crimea. The first X-ray observatory in space was named after Albert Einstein; the first large optical instrument to be launched into space was named after Edwin Hubble. Why was he the first astronomer to be given such an honour? Edwin Hubble died forty years ago; he had only a third of a century, with two world wars thrown in, for accomplishing his task in science.

The number of his published works is not especially large – only about seventy; the most significant of them appeared in the short period of five or six years in the middle and late 1920s.

It is precisely this work that Hubble is remembered for. It was Hubble who opened the world of galaxies for science when he proved that the nebulae outside the Milky Way are gigantic stellar systems often different from, but in many ways similar to, the galaxy which includes our Sun and its planets. However, the most important discovery was that of the red-shift in spectra of galaxies, which depends on distance from us and is perhaps the most revealing feature of the universe around us. The red-shift versus distance relation means that the world of galaxies is expanding. Therefore, the Universe was smaller in the past. By extrapolating back, contemporary physics has demonstrated that the primordial universe was different in everything – in the states of matter and radiation and in the rapidly evolving, violent development processes. What Hubble discovered is now known as the Big Bang, the primary explosion that gave birth to the universe as we know it. The explosive origin of the universe determined its subsequent evolution – the synthesis of chemical elements, the formation of galaxies and stars including our Sun and the planets, and ultimately the birth of living matter on at least one of them, the Earth; this gave rise to the human race with its desires, its passions and its rich history. This is why astronomers rank Edwin Hubble with Copernicus and Galileo Galilei.

Thus we have fulfilled the first criterion for prefaces.

Strange as it may seem, no monographs have been written about Hubble and even a sufficiently detailed biography remains to be published. Biographical data about him can be picked out from several obituaries and short entries, a few pages long, in reference volumes. Slightly more detailed data are included in a biographical essay on Hubble written by Nicholas Mayall and published by the American Academy of Natural Sciences. This essay is unfortunately too short.

It is not difficult to describe Hubble's work. His achievements are found either in his published papers or in the annual reports of the Mount Palomar and Mount Wilson observatories. To quote Alan Sandage who had worked with Edwin Hubble in the last years of his life; 'It seems to me that from the scientific standpoint, we know a great deal of what he did, and that was all documented in the records and in his publications. There is no question about the great things he did, but his personal life will be quite a bit more difficult to reconstruct'. An astronomer's life at an observatory is typically monotonous, regular and, outside professional interests, unexciting. It was not so with Hubble. He was always sensitive to events in the outside world and served not only science but his country too, in both world wars.

As far as we know, Hubble did not write reminiscences, as he believed that a scientist's life consists exclusively of his research work. One can hardly agree with that. The Russian physicist and historian of science Sergei Vavilov wrote that

The history of science cannot be limited to the development of ideas – to a similar extent it must be concerned with human beings, their peculiarities and talents and their dependence on the social conditions of their country and time... The life and work of pioneers in science are very important for progress in science and their biographies are a significant part of the history of science.

We agree with this attitude. We would like to tell the story not only of Hubble's work but also of his life, attempting to give this book a human dimension. It is not a simple task, though, for a variety of reasons. Hubble died many years ago. For new generations of astronomers in the USA, Hubble is a character from the history of science rather than a living memory. Only one or two still living Soviet astronomers actually saw Hubble, but they never had a chance to talk to him. Thus we have practically no living witnesses of Hubble's work in astronomy.

Another difficulty for the authors of the present book was that they had no opportunity to go through the Hubble archives, which are at the Huntington library in San Marino, a suburb of Pasadena where the Hubbles had lived. We did manage, though, to obtain copies of some significant documents from the archives and they are published for the first time in this book. We should note a peculiarity of Hubble's archives. The archives were set up and maintained by Mrs Hubble who not only adored but even idolised her husband. American researchers have noted that her devotion to the memory of her husband affected her selection of documents for the archives, which tend to present a somewhat idealised image of the late scientist. It was perhaps Mrs Hubble who depicted for Nicholas Mayall some events of their private life which have no independent confirmation and thus may be sentimental legends.

Though this biography is written in what we hoped would be a free rather than academic style, we have attempted a complete and thorough verification of all facts and events discussed in it. For each of them we have a recorded source. Unfortunately, in this way we may have automatically reproduced errors and misrepresentations of the originals when we lacked an opportunity to compare them with other sources.

In describing Hubble's research, we pay special attention to his main result: the discovery of the red-shift-distance relation. Its significance for our understanding of the universe is so great that a brief, or even detailed, account of it among his other results would be inappropriate. Therefore, we have dedicated a special section at the end of this book to the consequences of his great discovery.

The above outline is our answer to future criticisms of the book. We have greatly enjoyed collecting information on Hubble, often gleaning it in tiny bits. We should be happy if we have succeeded, at least to some extent, in depicting this highly attractive person, an outstanding scientist and a man of exceptional integrity. We were disappointed, as our project drew to a close, that we had failed to gather more biographical information on Hubble to share with readers. Robert Louis Stevenson expressed our feelings when he wrote: "To travel hopefully is a better thing than to arrive, and the true success is to labour." (From *El Dorado*.)

We should be satisfied with the thought that the book includes details that have been unknown to the American reading public. American researchers would definitely stand a better chance of writing a full biography of one of the greatest astronomers of the 20th century. They should not postpone this task for too long as the number of astronomers who can reminisce about Hubble, his work and his life is dwindling as the years go by.

We are grateful to all those who very kindly strived to help us in our project. They are the Russian astronomers P. Kholopov, P. Kulikovsky, N. Samus, A. Rastorguev and Yu. Efimov, bibliographer N. Lavrova, A. Dobrynin, the USSR Ambassador in the USA at the time when we started working on the book, Paul Hodge, Professor of Astronomy at the Washington State University, USA, Malcolm Longair, Astronomer Royal for Scotland, and many others. Our project was also warmly supported by Professor E. Kharadze.

We are especially indebted to Professor Kip Thorne of the California Institute of Technology. In fact, we regard him as our co-author. With the kind permission of the Huntington Library, he supplied us with documents from the Hubble Archives. He arranged for and conducted a taped interview with Alan Sandage from the Space Telescope Institute over long-distance telephone. He contacted Mrs Helen Lane, Edwin Hubble's sister, and asked her to write for us reminiscences which were essentially the only source of information about the early years of Hubble's life. We are extremely indebted to Mrs Lane and Alan Sandage for their kind assistance, without which the book would be incomplete.

When we received a warm personal letter from Mrs Lane in 1988 signed in slightly shaky handwriting – she was 89 at the time – we were pleased to learn that her family welcomed our intention to write a book about Edwin Hubble. We believe that it was quite unexpected for them that a book about the great American astronomer would be published in far-away Russia. It was encouraging to learn that a great-grandson of Mrs Lane was interested in astronomy and hoped to take on the profession of his illustrious relative. Let us hope with Mrs Lane that 'In the far future, perhaps, ... he will develop into an astronomer of his Uncle Edwin's caliber' and that the Hubble family will go down in history as one of those families in which talents in a particular field keep on appearing in several generations. Such were the musicians Bach in Germany, the mathematicians Bernoulli in Switzerland, the astronomers Struve in Russia and the USA, the biologists Thomas and Julian Huxley and the author Aldous Huxley; the latter was a personal friend of Hubble's.

This book is intended for a wide readership. We hope that both amateur and professional astronomers, particularly those who are in love with the history of their science, will find something new and entertaining in it. We shall welcome comments and remarks from our readers. We should be especially grateful if some astronomers can share with us unpublished facts and observations on Hubble's life and work that they retain from personal contact with Hubble or have inherited as legends coming down from their teachers.

PART ONE

Life and work

Choosing the way

The seventeenth century... Boatloads of new settlers kept coming to the New World from the continent of Europe. They had various reasons for leaving their countries – poverty, hopes for a better future, enterprising or adventurous temperaments, or persecution by the authorities. The astronomer's ancestor, Richard Hubble, an army officer, was among those who left Britain for America. It is not known precisely what made him leave the old country. Once Edwin hinted, though, that the reason was serious, merely saying that 'whenever there was trouble in England, one of the family left in a hurry'. Things were definitely in turmoil at the time King Charles I was executed.

Richard Hubble set up his family home in the state of Kentucky. He was successful and left five thousand dollars to each of his eleven children, which was a considerable sum of money at the time. One of Hubble's ancestors took part in the War of Independence on the side of the colonists. The grandfather of Edwin, Martin Jones Hubble, fought for the North during the Civil War. He later lost his official position for advocating clemency for the former enemy. Other members of the Hubble family fought on the Confederate side.

John Powell Hubble, a son of Martin's, married Virginia Lee James in 1880. Within a few years they had a son Henry and a daughter Lucy. On 20 November 1889, when the Hubbles stayed in the family house

in Marshfield, a small town in the state of Missouri, a boy was born
to them – Edwin Powell Hubble. Later the family steadily grew as Bill,
Virginia, Helen, Emma and Elizabeth were born. In fact, it was not
rare at the time to have many children in a family. The mother herself
was from a family of six children. This generation of Hubbles had very
different fates. Virginia, the family's favourite, died when she was still
a little girl. Lucy died when she was ninety-three. Two younger sisters
of Edwin, Helen and Elizabeth (Betsy) were alive at the time of writing
this book. Helen wrote in her reminiscences: 'I suppose that Henry and
Edwin would be considered to be of the intelligentsia and the rest of the
family was somewhat above the average, so far as standards of living,
ideals, ambitions and social graces were concerned'. But only Edwin
proved to be an outstanding person in this and subsequent generations
of the Hubble family.

John Hubble received an education in law and started a career as a
lawyer in Marshfield, but within two years his eyesight deteriorated and
he had to look for a more suitable occupation. He moved to Springfield
where his father had a successful insurance firm and specialised in fire
insurance. He was good at his job and the firm sent him to Chicago.
It was a new centre of the grain and cattle trades, of steel mills and
machine manufacturing plants, and of the famous Pullman truck plants
which were growing at a rate that was fantastic even by American
standards. John Hubble was of the opinion that a large industrial city
was no place for his children to live, so the family lived in the suburbs –
first Evanstone and then Wheaton where many years later Grote Reber
built the first parabolic radio-telescope near his home and observed the
Milky Way and the Sun. Now these formerly quiet suburbs are inside
the busy metropolitan area. John Hubble again suffered a deterioration
in his health and the firm gave him a new job in Louisville in the state of
Kentucky, where the family lived in the charming suburb of Shelbyville.

The family was well-off and they always had large comfortable houses
with a library, drawing room, dining room and a large foyer on the
ground floor, bedrooms and bathrooms on the first floor. The basement
floor was given to the children; they could rush around and play any
games there without fear of breaking something valuable. The house was
usually surrounded by a spacious courtyard for children's games and also
had a lawn where adults played tennis. Also in these grounds, there were
flower beds – nasturtiums, roses and asters – tended by their mother.

The ultimate authority in this happy world was their father, who was
unfailingly felicitous towards all his children. After he had returned

from his office, the dinner gong rang and at half past six the entire family would be at the dinner table. Their father was always very particular that everybody be at dinner on time. The atmosphere at dinner was somewhat formal but the children could always invite their friends if they had warned their mother beforehand. Their father was concerned that his children should get a proper upbringing and grow up as responsible citizens. He was a 'temperance' person and one of his worries was that his sons and their friends might yield to the temptation of drinking alcohol, which he strongly condemned. When Edwin went to study in England he had to vow to his father that he would not touch alcohol. His father was apprehensive that his sons might mix with bad company at the race track a mile from Wheaton.

Though the family employed several servants – cooks and parlourmaids – the children were not spoilt. Each of them had to make their own bed and to clean their own room. During summer vacations the father encouraged his sons to earn some pocket money. There were various ways for the boys to earn some money. Horse-driven wagons brought ice to the houses and the boys delivered the boxes of ice. Their father paid them for cutting the grass on the front lawn and in hot weather they could always sell some cold water to the workers building roads near the town.

The most exciting vacation job for Edwin was with a group of land surveyors who mapped the route of a railway line to be constructed in the woods in the region of the Great Lakes. The region was a wilderness at the time and one could have dangerous encounters there. It has been said that once Edwin was attacked by two robbers and was stabbed in the back. He knocked down one of the robbers and the other fled. When the season ended, Edwin arrived at a railway station with the work party but found that it was deserted since the train timetable had been changed and no train was due for a long time. The land surveyors decided not to wait for the train and went on foot through the woods for three days without any food.

Perhaps it is a myth, but Edwin was said to have returned from this adventure as a grown man.

The money Edwin and his friends earned gave them a welcome feeling of independence. Money in their pockets gave them freedom to do a lot of pleasurable things – go to a fair and buy whatever they fancied, pay for their favourite tunes to be played by a nickelodeon, buy ice-cream cones for girls they liked.

Life at home had many pleasant sides, too. The children often congregated around a large table to do their homework together. They especially liked weekends even though they were obliged to attend church services and Sunday school lessons. Their father was deeply religious and each Sunday morning the entire family and house guests solemnly set out for the church. After that, the children were free to go swimming or riding hay wagons in summer or sledge riding in winter. There were numerous parlour games to play indoors if the weather was bad. They often played a word game in which one had to add a letter to a word so as to avoid being the one on whom it ended. To win, one had to be inventive and look up words in a dictionary. In the evenings they had family concerts. Their father played the violin, Lucy, who had a considerable musical talent, played the piano and Edwin and Bill played mandolins. The smaller children seldom tired of listening to them.

Edwin loved playing with his younger sisters. Betsy's earliest memory is of how Edwin took her to a circus on his shoulders. When Betsy was twelve, she visited her brother when he was already working at the Yerkes Observatory. Telling her to be very quiet, Edwin took her to a telescope and she held her breath, looking for the first time at the heavens through the eye-piece. Once Edwin took his two younger sisters to a performance of *The Blue Bird* by Maeterlinck and it was difficult to say who enjoyed it more – the young man or his charges.

Family life had its dramas, too. When Edwin was six Virginia was just two. She was a nice girl and Edwin and Bill sincerely loved her. But she often destroyed the toy houses, fortresses and bridges the boys built and once they subjected her to a rather painful and severe punishment. Within a few weeks Virginia fell ill and soon died. The brothers thought it was their fault and Edwin had a nervous breakdown. If the parents had been less sensitive and caring there could well have been another family tragedy.

Their mother was especially attentive towards the children. Any dispute, large or small, was quietly resolved by her and she always succeeded in maintaining the spiritual harmony of the family. Mrs Lane remembered that she had seen her mother angry only once, when the eldest children, Henry and Lucy, had pinched nuts from a dessert table specially set for a party. His mother was an attractive, statuesque woman and Edwin inherited her graceful bearing and elegant manners.

Like most boys, Edwin went in for sports. In the last years of high school he grew to six foot three and won prizes for basketball and pole vaulting.

Thus Edwin had a normal childhood in a middle-class family, in a kind and well-ordered household. As a grown-up, Edwin said that the family discipline had prepared him for much. There was perhaps one feature that distinguished him among other students, namely that he learned to read and write very early: when his elder brother and sister went to school he was deeply sorry that he was too young to go to school with them. Edwin loved reading and he read numerous adventure books, which were favourite reading for boys of his age at the turn of the century. His favourites were novels by Jules Verne and H. Rider Haggard, particularly *King Solomon's Mines*.

Reading about famous people in whatever field they have distinguished themselves – arts, science or writing – one invariably wishes to know when they started to show interest in their field, when they chose their careers. Edwin apparently became fascinated with astronomy at a very early age, under the influence of his maternal grandfather, William Henderson James. This grandfather was definitely an exceptional person. He was educated as a medical doctor in Virginia but the Gold Rush took him to California, where he spent his youth. Here he married Edwin's grandmother and then returned with her to Missouri. He was enthralled by astronomy. Mrs Lane recalled that their grandfather 'rigged up a telescope that simply charmed Edwin to such extent that he requested that in place of his eighth-year birthday party, he preferred to be allowed to stay up late to look through the instrument to his heart's content. The wish was granted'.

When Edwin was twelve he wrote such an interesting letter to his grandfather answering questions about Mars that it was published in a Springfield newspaper.

At the age of fourteen, Edwin had surgery for appendicitis. Today this is regarded as a simple operation and patients do not have to stay in bed as long as they used to. But Edwin liked being bed-ridden – he had plenty of time to read about stars.

Memoirs are not the most reliable historical source, since human memory is imperfect. Edwin's friend, the famous English author, Aldous Huxley, also mentioned this early fascination with astronomy in his essay *Stars and Man*. The details were different, though. For instance, eight-year-old Edwin was described as dreaming about watching meteor showers and his letter to his grandfather was described as a treatise on Mars. Perhaps that was how Hubble himself told these stories to the author of the essay.

Edwin graduated from high school in 1906. The studies were apparently too easy for him and he never made visible efforts in school. At the graduation ceremony, the school principal addressed him with the following words: 'Edwin Hubble, I have watched you for four years and I have never seen you study for ten minutes'. The principal made an ominous pause and continued, 'Here is a scholarship to the University of Chicago'. At the age of sixteen Edwin entered the University of Chicago, which had been founded in 1892 and was one of the ten best universities in America at the time. Its faculty at the turn of the century included the astronomer Forrest Moulton who suggested a theory on the formation of the Solar System, and the famous physicists Albert Michelson and Robert Millikan. It has been said that Edwin even worked for a time as an assistant in Millikan's laboratory. He was particularly drawn to astronomy and mathematics.

Edwin did not leave any notes about his student years. He lived far from his family, so his sisters can remember only very little about this time of his life. They, that is, Mrs Lucy Wasson and Mrs Helen Lane, can only recall that he continued playing basketball. Once his team played against the Wisconsin University team which included his brother Bill. The university football coach Alonzo Stagg wanted Edwin on his team and asked John Hubble to allow his son to join it. The father thought, however, that football was too violent a game and refused to be persuaded. The football-playing friends whom Edwin invited to his family for week-ends failed to impress his father. Edwin's father's favourite game was baseball, and the cunning Edwin then started talking at length about injuries suffered by baseball players. His father declared then that Edwin had better stop playing this game, too. Father's word was law in the family. Edwin switched to boxing, thinking bitterly that his father had ruined his life (this opinion was ventured by Lucy).

Edwin proved to be so good at boxing that the coach thought of making him a professional. Luckily, things never went that far.

Edwin dreamt of winning the Rhodes scholarship (named after Cecil Rhodes, whose name until recently featured on the map of Africa – Rhodesia). This annual scholarship of 200–300 pounds sterling was quite enough for a young single gentleman to lead a comfortable life in England at the time. Professor Ernest Rutherford even suggested a similar salary at his laboratory to Niels Bohr, who was already married. The scholarship was awarded to the best unmarried students of 19–25 years of age who were healthy, excelled in sports and had unblemished character – the latter was of the utmost importance. Edwin, unfortunately, was not always on his best behaviour, and once, with his club-mates, engaged in throwing raw eggs out of their windows at theology students' suits being returned from the cleaners. Plastic bags were not used for packing things at the time and the students took very careful aim.

The incident was overlooked, fortunately, and in September of 1910 Edwin and a number of other young Americans went overseas on a two-year scholarship which was subsequently extended by a further year. In England, Hubble's life underwent its first sharp change. Though he majored in natural sciences in Chicago, in England he had to take law as his subject. Hubble never explained the reasons for this change. Probably, it was the scholarship supervisors who decided it for him. Mrs Wasson related a few facts about this period to a reporter of the town paper in Alexandria where she lived at the time.

The reporter came to interview the sister of a famous man mentioned alongside Copernicus, Galileo, Kepler, Newton, Herschel and Einstein in an article on 'Pioneers in man's search for the Universe' published in the highly popular monthly, *National Geographic*. Talking about her brother, Mrs Wasson remarked that 'he was not assigned to Cambridge, where he could have taken advanced science. Instead, he attended Queen's College in Oxford, where he studied international law'. The universities of Cambridge and Oxford are famous in that they educate the elite of the nation. Over sixty Nobel prize winners have graduated from Cambridge, most British prime ministers after World War I studied in Oxford and most British judges were educated in these universities. At ancient Queen's College, Edwin mixed with future statesmen, sons of the nobility and future prominent authors. Students received a liberal education in Oxford but in addition they learned how to communicate, how to behave in social circles, how to enjoy culture and sports. It was, of course, through his Oxford law studies that Edwin developed his natural talent for precise, expressive, convincing and dramatic presentation of his ideas. He started his book-collecting while in Oxford. A few years after his death his sisters presented El Paso University with two old books bought by him in Oxford – a Latin volume of the 16th century and works of Thomas Robert Malthus.

It was in Oxford that Edwin learned the exceptionally reserved and dignified manner of behaviour that distinguished him throughout his life. There also he learned to smoke a pipe, which became almost his emblem, making him look older and somehow more scholarly. Alan Sandage, who knew Hubble in the last few years of his life, remembered his 'absolute spiritual power, moral integrity, and some air of superiority... In England he would be a genuine nobleman, one of the High Society in all respects'. In the best Oxonian traditions he devoted a great deal of time to sports –

boxing, rowing and athletics. Edwin also wanted to know Europe better and in his two summer trips to Germany he travelled over 2000 miles, three quarters of them by bicycle.

Having graduated with a BA degree in law, Edwin returned home to Louisville in the summer of 1913. His father did not live to see him return, he died in January of the same year. Edwin and the older children were independent by that time, while the youngest, Elizabeth, was just eight years old.

Most biographies of Hubble state that he practiced law for a year. But a few years ago Kip Thorne and Alan Sandage received information from a historian in Kentucky who claimed he had documents proving that Hubble had taught at a school and had been a coach of the school basketball team. Mrs Lane recalls also that Edwin did German translations for some firms. She claims that he never practiced law. She closely observed his life that year in Louisville. She remembered some small details like her mother treating Edwin's friends to tea and exquisite cinnamon rolls. Edwin and his friend Walter Stanley Campbell played mandolins and one of them sang. Edwin enjoyed the quiet family life that he had been missing in his student years.

Another of Hubble's sisters, Mrs James, finally resolved this question when she contacted the Kentucky authorities and received an official answer that, after checking all their records, they could 'find no evidence that Edwin Powell Hubble was ever admitted to the practice of law in Kentucky'. This is the end of one of the myths about Hubble's life.

Edwin was not satisfied with what he was doing. Astronomy still attracted him. He realised apparently that the three years in England had been a step away from his preordained path. He said once to a newspaper interviewer that 'Astronomy is like ministry. You need a calling. After practicing law for a year in Louisville, I got the calling'. Biographers usually quote his words that he abandoned law for astronomy. He claimed that he 'knew that even if I were second-rate or third-rate, it was astronomy that mattered ... ' It is difficult to judge whether Hubble really said something to that effect or whether it is one of the myths about him. If he did say that, it would mean that he was reluctant to admit that the Rhodes scholarship was wasted on him, and he pretended to have tried making use of his studies for a whole year. It is quite possible, though, that he never said anything of the sort and someone who wanted his biography to be prettified invented those words.

The final choice was at last made. Hubble returned to the University of Chicago to work on a PhD thesis at the Yerkes observatory. The observatory funded by the Chicago railroad magnate Charles Yerkes soon gained recognition among astronomers of the entire world. It boasted a 40-inch refractor that is still the largest telescope of this type. A 24-inch reflector was built later at the observatory. Hubble's supervisor was the director of the observatory, Professor Edwin Frost, who specialised in astrospectroscopy.

In August of 1914, Hubble visited the 17th Meeting of the American Astronomical Society in Evanstone, where he met personally, for the first time, many American astronomers. He was elected to the Society together with other astronomers who would be well-known in the future, such as F. G. Pease, H. H. Plackett, A. van Maanen, Loise Jenkins (who compiled famous catalogues of stellar parallaxes and bright stars), and many others. The participants of the Meeting were photographed together, with the Society President Edward Pickering at the centre, surrounded by well-known astronomers. The tall figure of young Hubble is seen in the right-hand corner of the photograph.

Perhaps, it was at this meeting that Hubble learned for the first time that Slipher had determined the radial velocities of thirteen nebulae. The measurement results were quite unexpected. In contrast to stars, many nebulae had very high velocities, and many of them were receding. These observations probably made a lasting impression on Hubble.

Hubble started his work at the observatory by photographing with the reflector. His first published paper was concerned with stars exhibiting appreciable proper motion. When Hubble used a blink comparator to compare his negatives with those made by his colleagues approximately ten years earlier, he found fairly large displacements – from 0.2″ to 1.5″ a year – for twelve stars. Four of these stars were of magnitude 15^m or fainter. He wrote:

So far as I am aware, these are the faintest stars in which appreciable motion has been found... In view of the small number of fields examined ... it is reasonable to suppose that considerable numbers of such faint stars exist in the immediate neighborhood of our Sun.

Indeed, all these numerous stars proved to be dwarfs of late-type spectral classes located at distances of several tens of parsecs from the Sun.

Using Barnard's plates, Hubble discovered twelve unknown variable stars. He failed to complete this study because of World War I; he managed to publish a short report on the new variable stars only in 1920 when he was already on the staff of the Mount Wilson Observatory.

Later Hubble became interested in the cometary nebula NGC 2261 with the well-known irregular variable star R Monocerotis. When he compared the photograph he made on 8 March 1916 with a plate exposed by Jordan in 1908, he realised that the structure of the nebula had changed. He compared his results with five negatives photographed at different times at the Yerkes and Lick observatories and he managed to obtain a copy of the photograph made by Isaac Roberts in England. The variability of the nebula was verified. In spring of the next year Hubble found new changes in the nebula. He suggested that the most probable explanation of these changes was motion of portions of the nebulosity relative to the nebula as a whole and the ejection of matter from its nucleus. This nebula kept drawing Hubble's attention. He continued studying it for several years and studied it again thirty years later, at the end of his life.

But neither stars nor cometary nebulae were the subject of his doctoral dissertation. It was entitled *Photographic Investigations of Faint Nebulae.*

About two thousand nebulae had been discovered by that time but their nature remained unknown. The term 'nebula' was applied to various objects – the gaseous formations in our stellar system, diffuse and planetary nebulae, and the distant galaxies. To observers, they looked alike in that their photographic images could not be resolved into stars. Some astronomers had guessed already that the spiral nebulae with unusually high radial velocities and immeasurably small proper motions were outside our galaxy. The nature of numerous small and faint galaxies was quite unclear. They had to be studied statistically and the first step was to conduct systematic observations with a sufficiently powerful telescope.

Hubble photographed seven fields far from the Milky Way, using a reflector telescope. A careful analysis of the photographs revealed 512 new nebulae in addition to the 76 detected before. He measured their coordinates and described them, noting their shape, brightness and size. In one field the nebulae were apparently grouped together: a region of the size of the full Moon contained 75 objects out of 186 recorded in the entire field. Hubble concluded: 'Suppose them to be extra-sidereal and perhaps we see clusters of galaxies; suppose them within our system, their nature becomes a mystery'.

After making this conclusion, Hubble presented some shaky proof that the spiral nebulae were indeed distant stellar systems or galaxies. Although the initial data and calculation methods were questionable, the results still tended to show that 'the spirals are stellar systems at distances to be measured often in millions of light years'. Hubble noted that the same conclusion followed from the results on the motion of the Sun with respect to the spiral nebulae, derived from their radial velocities by Truman, Young and Harper.

For a few years after finishing his thesis, Hubble discontinued his studies of the extragalactic nebulae.

After World War I, he shifted his attention to diffuse nebulae in our galaxy. But later he returned to what was his absorbing scientific interest until his last days – the amazing world of galaxies which he effectively opened to mankind. His mentor Edwin Frost, who died in 1935, had reason to be proud of his pupil.

The fatal gun-shot in Sarajevo on 28 June 1914 that killed Archduke Franz Ferdinand, the heir to the throne of the Austro-Hungarian Empire, rang as the warning shot for World War I. For $51\frac{1}{2}$ months, millions of Frenchmen, Englishmen, Russians and others were killing one another. The Archduke's death was not, of course, the real reason for the war. World War I was deeply rooted in the nature of the countries that were to be locked into the bloody fight; it stemmed from contradictions between the opposing groups, their competition for influence, for raw materials and markets, for world domination.

The main struggles raged on the Western front between Germany and England and France, and on the Eastern front between Germany and Russia. 'You will be back before the trees have shed their leaves' were the words with which Kaiser Wilhelm II sent his troops into battle. He was not the only leader to be so mistaken. The bloody conflict cost the lives of 10 million, maimed 20 million and separated 76 million people from their families and work. Both sides were locked in endless, fruitless offensives and retreats, never leaving the hated trenches. Barbara Tuchman wrote in her book *The Guns of August* that

nothing could give dignity or sense to monstrous offensives in which thousands and hundreds of thousands were killed to gain ten yards and exchange one wet-bottomed trench for another. When every autumn people said it could not last through another winter and every spring there was still no end in sight, only the hope that out of it all some good would accrue to mankind, kept men and nations fighting.

Until spring 1917 the USA kept out of the war in Europe, but on 6 April President Woodrow Wilson proclaimed a state of war against Germany and its allies.

At that moment, astronomy, the most peaceful of sciences, was on the brink of a great event. The manufacturers were finishing the reflector for the 100-inch telescope of the Mount Wilson Observatory which was to make the greatest contributions to progress in astronomy for many years to come. Edwin Hubble's research was to be linked to this instrument. The observatory director, George Ellery Hale, decided to invite the best American astronomers who were capable of working on the most important problems in astronomy. Before, Hale had been the director of the Yerkes observatory and knew Hubble from his Chicago days. He probably recognised immediately Hubble's dedication to science, his rare determination and stamina.

In spring 1917 Hubble was making a last-minute dash to finish his thesis. One day, after spending a sleepless night writing his thesis and the next day passing an oral examination, Hubble made a decision that sharply changed his life. Director Hale received a short telegram: 'Regret cannot accept your invitation. I am off to the war'. Hubble joined the American Army as a volunteer.

It is not easy to understand the reasons for this decisive step. Hubble's biographers have given no clear indications here. He was no longer a careless youth. He was a gentleman of twenty-eight who had graduated from the University of Chicago, continued his education in Oxford and finally obtained a PhD degree. He was offered an exciting opportunity to do research at the observatory where the largest telescope in the world would soon become operational. It seems very unlikely that Hubble's decision to volunteer for the army was impulsive. He had lived in England, he loved this country and he decided to help it to fight its enemies. Quite a few young Americans from the academic community joined the army at that time – there were 325 from Harvard, 187 from Yale and 34 from Chicago. Hubble was sent to the First Officers Training Camp in Fort Sheridan, Illinois, and there he was taught his third profession, after science and law – the military one.

The USA had not had a big regular army in peacetime. Now they had to organise one. On 5 August 1917 the 86th Black Hawk division of the National Guard was established. Its emblem featured a black hawk against a three-cornered red shield. In late August the officers under training at the First Camp were ordered to join the officer corps of the division. Captain Edwin Hubble was made commander of the 2nd battalion of the 343rd infantry. In December 1918 he was promoted to major. The officers were busy training newly-drafted soldiers from Illinois and Wisconsin. It was still a long way to go to the European theatre of operations. Only in August of the next year was the division brought to New York to board military transports for Europe.

World War I ended in late 1918. The Russian government signed a separate peace treaty with Germany, thus weakening the forces of the Entente. The arrival of the American expeditionary forces meant a worsening of the situation for the German high command. It made a last-ditch attempt to beat the allies on the Western Front before American might was brought into play. For 118 days in March, April and May, the German armies staged an offensive, with only short-lived success. But in July 1918 the Entente, reinforced by American supplies and troops, started an advance that ended in Germany's capitulation in November.

American troops were pouring into Europe. In May 1917 there were only 1300 of them but, by the end of the war, over two million had arrived. Hubble's division boarded troop ships on 8–9 September 1918 and reached England in about a fortnight. Dozens of German submarines were hunting American transports in the Atlantic, but by autumn 1918 the danger was greatly diminished. Convoys were securely guarded by cruisers and frigates and hundreds of airplanes and balloons were on the look-out for submarines near the English shores. Hubble's division safely reached Liverpool and on 23 September the troops disembarked at Le Havre and Cherbourg in France. From there, the division moved southwards towards Bordeaux. Only six weeks were left until the end of the war. Like fifteen other American divisions, the Black Hawk was too late to get to the battle front. It remained near Bordeaux until 8 November, when it was moved towards Le Mans.

We do not know much about Hubble's military service. Probably, nothing exceptional happened, it was just routine service. Hubble did not experience any discomfort in the army. He liked the strict army order, the discipline, the straightforward relations between men. He was not, however, an admirer of army drill for its own sake. He never lost his sense of humour. There is a funny story from Hubble's army record. Once he was riding a bicycle on the drill ground. When he saw a general coming his way Hubble stopped, saluted and said, 'Good morning, general, nice day, sir'. 'Major, could it be that you do not know my orders concerning the procedure to be taken by an officer meeting me?' barked the general. The orders were, of course, to give your rank and name and to state what you were doing. Hubble executed the drill precisely. He stood to order, smartly saluted the general, jumped on the bicycle and reported, riding away, 'Sir, Major Hubble, 86th Infantry, getting on his bicycle and riding away'.

By October 1918 the imminent defeat of Germany became certain. Its allies capitulated one by one: first Bulgaria signed an armistice, then Turkey left the war, and then the Austro-Hungarian empire capitulated. On 8 November, the German delegation arrived at the Compienne forest and three days later signed the capitulation document dictated by Maréchal Foch. The Command of the American Expeditionary Force in Europe issued an order to stop hostilities the same day:

Monday morning: In accordance with the terms of the Armistice, hostilities on the fronts of the American Armies were suspended at eleven o'clock this morning.

Though Hubble's division was never in battle it suffered at least one casualty. Major Hubble was concussed and his right arm injured by an accidental grenade explosion (unless this is yet another Hubble myth). It took a long time to bring back the American troops, even though they were all anxious to get home as soon as possible. Units of Hubble's division started going back in December but it was only in early August of the following year that the last of them returned home. By that time Hubble was no longer with his regiment. First he was posted to the court-martial branch because of his legal education. Later he was appointed administrator of a group of American officers admitted to the universities of Oxford, Cambridge and Wales for short courses before their returning to the USA. After spending almost a year in France and England, Hubble came home in late August 1919. He brought with him his tin hat, the crossed rifles of the 343rd Infantry, the gold maple leaves of a major and a German trench dagger – 'to cut the pages of French books'.

Hubble was mustered out in San Francisco and hurried to Pasadena to take up Hale's offer. The vacancy was still open and in September Hubble was put on the staff of the Mount Wilson Observatory. He was at the beginning of the most important stage in his career, into which he entered as a budding astronomer and emerged as an outstanding scientist of the 20th century.

On the threshold of great achievements

The Mount Wilson Observatory was founded in the spring of 1904 by the outstanding solar research astronomer George Elery Hale. Hale is remembered in the history of astronomy mostly as an excellent science manager who saw far into the future, was full of energy and initiative, untiring in the pursuit of a goal and exceptionally successful. His brain was teaming with projects, each one greater than the one before. He was the man who persuaded C. T. Yerkes to provide the money for the one-metre refractor and for the construction of the observatory near Chicago.

Hale realised that successful solar observations required huge instruments installed at locations where days of clear weather were especially numerous. Southern California was just such a place. In 1901, Andrew Carnegie, who pioneered the Bessemer process of steel manufacture and became enormously rich, retired from business. Carnegie gave considerable sums to charities, paying for the organisation of numerous libraries; in 1902 he founded the Carnegie Institution in Washington. Hale accidentally found out about this and immediately realised that it would be stupid to miss this golden opportunity. Several months later the Carnegie Institution paid for moving the solar telescope from the Yerkes Observatory to the summit of Mount Wilson in California. The observation conditions proved to be exceptionally good. Clear atmosphere reigned two hundred days a year, and observations could be conducted for parts of a day for another one hundred days. Abbot's measurements confirmed the high transparency of the atmosphere, and Barnard was able to take excellent photographs of the southern Milky Way, with stellar clouds and extended nebulae which were previously unknown.

Originally, the Mount Wilson Observatory was known as the solar observatory; its main objective was to analyse the Sun 'as a typical star, in connection with the study of stellar evolution'. In fact, Hale, who became the director of the new observatory, was dreaming of a day when huge stellar telescopes would be installed on Mount Wilson. In 1908, the then world's largest 60-inch reflector became operational at the observatory. While a mule train was pulling a carriage with a mirror for this instrument along a road climbing towards the observatory, Hale was informed that a French mirror glass company in St Gobain had already manufactured a 100-inch glass disc that Hale had ordered. The money to pay for the mirror was provided by the Los Angeles tycoon John Hooker. The Carnegie fund responded by allotting money for the construction of the telescope itself.

During this period, Hale's interest turned to spiral nebulae. In 1907 Hale wrote: 'A 100-inch reflector, with focal length of 50 feet, should be capable of photographing an immense number of those objects to excellent advantage'. The true nature of the spirals being unknown at the time, Hale planned that solution of the problem of planetary system formation would be the main objective of the new telescope. Hale continued: 'In view of the work of Chamberlin and Moulton on the nebular hypothesis, and the theoretical study of spiral nebulae ... , such

observational evidence as the new instrument may afford should prove
of the greatest value'. The 100-inch reflector did play a tremendously
important role in astronomy, although not in the way that Hale projected
for it.

The pattern of life at the Mount Wilson Observatory had already
settled by the time Hubble arrived. The time when astronomers had
to rough it out in the 'Monastery' – a small one-story building on the
mountain slope, with the office, a library, and living quarters – was long
past. A two-storey building on 813 Santa Barbara Street, with a basement,
study rooms, library and administrative offices had been designed by the
architect M. Hunt. The astronomers were especially happy that they
were able at last to put their books and periodicals in proper order. The
library boasted a portrait of one of the pioneers of astrospectroscopy, Sir
William Huggins. A large oil portrait of John Hooker was added later,
after Hooker's death in 1911. Implementing George Hale's plans, optical
works, a physics laboratory and other buildings were later erected on the
premises.

Cars were at that time regularly commuting between Pasadena and
the observatory along a good road, so that it took slightly more than
two hours for observers to reach the summit. But the main event was the
beginning of operations of the 100-inch reflector in November 1917.

Hubble joined the group which photographed nebulae; the group
included J. C. Duncan, F. G. Pease and R. F. Sanford. Here he met
Milton Humason who had just joined the research staff. Humason, son
of a Californian banker, had come to work at the observatory as a janitor
two years earlier. He had decided at the age of fourteen that he had
had enough schooling to last him for the rest of his life. The other two
janitors stayed in that position to the end of their days but the inquisitive
Humason, who learned how to drive a team of mules, how to take care
of the clock mechanisms and to help in the photographic laboratory,
attracted the attention of the scientists. He soon started reading about
astronomy in earnest, was promoted to the position of night assistant,
and later became an outstanding observer – one of the few with whom
Hubble later collaborated.

Humason remembered:

My own first meeting with Hubble occurred when he was just beginning observations on Mount Wilson. I received a vivid impression of the man that night that
has remained with me over the years. He was photographing at the Newtonian
focus of the 60-inch, standing while he did his guiding. His tall, vigorous figure,
pipe in mouth, was clearly outlined against the sky. A brisk wind whipped his

military trench coat around his body and occasionally blew sparks from his pipe into the darkness of the dome. 'Seeing' that night was rated extremely poor on our Mount Wilson scale, but when Hubble came back from developing his plate in the dark room he was jubilant. 'If this is a sample of poor seeing conditions,' he said, 'I shall always be able to get usable photographs with the Mount Wilson instruments'. The confidence and enthusiasm which he showed on that night were typical of the way he approached all his problems. He was sure of himself – of what he wanted to do, and of how to do it.

Hale was not mistaken in his choice. The new member was vigorous and full of plans. Hubble was working on virtually all instruments: the 60- and 100-inch reflectors and, during the initial phase, the 10-inch astrograph. The reports of the observatory frequently mention the hundreds of photographic plates that he obtained, often after exposures of 4–5 hours. He photographed one area in Ophiuchus for 19 hours, exposing the plate three nights in a row. Actually, D. Osterbrock noticed many years later that 'Hubble was technically a rather poor observer, as his old Mount Wilson photographic plates show, but he had tremendous drive and creative insight'. High-quality negatives are a result of the art of observation (not everyone is gifted with this ability) and of the choice of a night when the stellar images are not blurred but appear as well-defined dots. It is not easy to indicate where Hubble's imperfections were, in the ability to observe or in the willpower to stop observations under poor atmospheric conditions. Presumably, Hubble lacked the latter, in which case there would be a precious few spectacular photographs among numerous plates: Mount Wilson was never regarded as a place of exceptionally quiet atmosphere.

Hubble's first objective was to study the bright and dark nebulae along the Milky Way. Many years later Mayall wrote:

His knowledge of individual nebulae was encyclopedic. The 100-odd Messier objects were as familiar to him as the alphabet. He knew literally hundreds of NGC objects in sufficient detail to recall their structure, their relationship to neighboring ones as pairs, multiples, or clusters... He knew the Milky Way, with its complex structure of bright and dark nebulosities, star clusters, planetaries, and nebulous stars, as thoroughly as any port pilot threading his way through a tortuous system of channels, bars, and buoys. On one occasion on Mount Wilson when a tenderfoot from Berkeley was trying, unsuccessfully, to find an object with the 60-inch, Hubble entered the dome, recognized the military fluid situation, sighted along the tube without leaving the floor, and said: 'The declination is plus five degrees'.

Hubble fixed several wide-angle cameras to the 10-inch astrograph, in order to photograph large areas of the sky. The photographs defined large-scale details of the Milky Way. Several large dark nebulae were found in this way far from the middle line of the Milky Way, and Hubble and Duncan were able to locate one such nebula 37 degrees away from it. This might seem to justify detailed description but for some reason Hubble never published anything about this observation.

In fact, this was not the only case. Hubble regularly used large reflectors to photograph a variable nebula in Monocerotis which he had already started to follow while at the Yerkes Observatory; he also observed other objects. He dropped some of them within three years but kept following others continually. Nevertheless, he did not publish even a short note. Although the observatory's reports were typically precise and reliable, and one of them mentioned three new globular clusters found by Hubble, we still do not know which they were. They are absent from his cluster catalogues.

Hubble's first paper after his return from the army was a short note about a group of faint nebulae, jointly numbered as NGC 1499, scattered in Perseus over an area of several lunar discs. Hubble discovered on photographs taken with an objective prism that their spectra contained emission lines of hydrogen, helium and oxygen. The same photographs told him that the already known small nebula IC 2003 is a planetary nebula. Its spectrum manifested especially bright characteristic lines of 'nebulium' – a putative new element whose nature was understood only much later.

A survey of the sky using the objective prism brought new discoveries. Hubble detected twelve planetary nebulae. Half of them had already been spotted by other observers but Hubble was able to establish their true nature. Six faint small nebulae, whose central stars were not found even by large telescopes, proved to be of special interest. Now such compact objects are regarded as planetary nebulae at the earliest stages of their formation. Humason also discovered two small nebulae, and Hubble studied them at the focus of the 60-inch reflector. Apparently, this was the first case in which these two outstanding observers combined their efforts.

Photographs through the objective prism showed Hubble not only galactic emission nebulae. He also mentioned broad, bright lines in several objects which many years later were given the name of the astronomer who studied them: Seyfert galaxies. Hubble was probably the first to discover emission in the spectrum in one of them, the NGC 4151 galaxy.

In May 1922, Hubble presented to the *Astrophysical Journal* an excellent paper entitled 'General Study of Diffuse Galactic Nebulae'. This work has not lost its importance even now, since the progress of science has only supported the view that Hubble had advanced many years ago.

First of all, Hubble clearly classified all nebulae into two types: galactic nebulae connected with the Milky Way band, and extragalactic nebulae at high latitudes, which are absent in the plane of our stellar system. It had not yet been reliably established that the latter objects lie beyond our Galaxy, that they are stellar systems similar to ours but removed from it by tremendous distances. Hubble was able to undertake their study later.

Two types of galactic nebulae exist: planetary nebulae and diffuse nebulae whose shape is not as regular and well defined. In their turn, diffuse nebulae may be dark, resembling a sort of void against the background of the sky, or they may be bright, sometimes resembling an elegant team of tenuous clouds. Navigators of southern seas knew quite well one of these dark spots in the bright Milky Way: the Coal Sack. Hubble was the first to show that the positions of bright diffuse nebulae tend to concentrate in the Milky Way and in the great circle inclined to the Milky Way at an angle of about 20 degrees. Also concentrated along this circle, known since 1879 as Gould's belt, were bright blue stars. The dark nebulae observed by Hubble also manifested this tendency.

Having studied the spectra of about 60 bright nebulae, Hubble discovered that the spectra of about half of them were continuous. These were the nebulae that were mostly grouped around Gould's belt. The others, mostly located in the Milky Way, displayed emission lines in their spectra. But these were not planetary nebulae, neither in shape nor in spectral features. The hydrogen emission line H_β in these nebulae was more intense than the nebulium lines.

Hubble was able to find stars that were obviously connected with the light diffuse nebulae, but their characteristics were different. In continuous-spectrum nebulae, these were stars of later spectral classes, beginning with B1, and in emission nebulae these were higher temperature objects, of O–B0 classes.

Definite relation between the spectra of the nebulae and of the associated stars suggests that the source of luminosity of the nebulae is the radiation from those stars. According to this view, the nebulosity has no intrinsic luminosity but either is excited to emission by light from a star of earlier type or merely reflects light from a star of later type.

This was the main conclusion that Hubble drew from his study.

In his next paper, Hubble was able to test quantitatively the hypothesis that stars were indeed the source of light from nebulae with continuous spectra. Slipher and Hertzsprung carried out such tests on several objects before Hubble. However, Hubble collected more than 80 photographs of nebulae, obtained both on the largest instruments – 60- and 100-inch reflectors – and with small cameras. It is not difficult to show that under identical photographing conditions, the limiting size to which a nebula can be traced and the apparent magnitude of a star seen in this nebula are related in a certain way. This was indeed found to be the case. Hubble wrote:

Diffuse nebulae derive their luminosity from involved or neighboring stars and that they re-emit at each point exactly the amount of light radiation which they receive from the stars. Where stars of sufficient brightness are lacking in the neighborhood, or, if present, are not properly situated to illuminate the nebula as seen from the Earth, the clouds of material present themselves as dark nebulosity.

Those diffuse nebulae whose light is strongly absorbed by cosmic dust inevitably deviate from the relation formulated by Hubble. A feature of even greater interest was that the planetary nebulae emitted at a hundred times greater intensity than the simple theory predicted. This fact signified that a nebula was receiving from its star invisible ultraviolet radiation which was then re-emitted in the visible band detectable by the astronomer. This was the correct conclusion that Hubble drew from his observations. But he also considered it possible that the emission of a planetary nebula can be excited by a flow of corpuscles emitted by the star.

The problem on which Hubble worked during his first years at Mount Wilson was thus solved. He was facing a new, immeasurably vaster field of research: extragalactic nebulae.

When Hubble was studying galactic nebulae, he was undoubtedly thinking about nebulae located outside the Milky Way; in fact, they attracted him before the war started. These objects had been photographed at the observatory for some years already, especially on spring nights when the Milky Way stretches along the horizon. The photographs obtained with the largest instruments presented to the observers a world of most various and at the same time surprising common forms. The first problem was to classify the structures of these objects and to identify their basic types.

Hubble began his classification of both galactic and extragalactic nebulae at the same time. His analysis of a large number of plates allowed him to single out four types of extragalactic nebulae: spiral, elongated (which included spindle-shaped and ovate nebulae), globular and irregular nebulae; the last group subsumed the nebulae that could not be classified under the other three groups.

This classification was a certain compromise between the ones suggested before Hubble. For instance, Curtis tended to believe that there were only three types of nebulae: planetary, diffuse and spiral. However, photographs taken on large reflectors made it clear that not all nebulae visible outside the Milky Way possessed a spiral structure. Some of them did not manifest even the slightest attributes of spiral arms; the photographs showed them as elliptic objects with brightness decreasing towards the edges. One example was the nearest neighbour of the spiral Andromeda nebula, the almost circular M32 nebula. Some of the nebulae were definitely of irregular shape.

Another extreme was an excessively detailed or descriptive classification without clearly defined criteria. For example, the scheme suggested a long time before by Max Wolf consisted of 23 types; it was not very easy to assign a nebula to a specific type. Reynolds' classification was also rather fuzzy.

The American scientists submitted Hubble's classification to the commission on nebulae of the International Astronomical Union. Only two corrections were made: one superfluous word was removed, 'elongated', and two subdivisions, 'spin-shaped' and 'oval', were proposed as independent types. Nevertheless, the incumbent president of the commission, the French scientist Bigourdan simply ignored the work of the then unknown American astronomer and suggested in his report his own, already published classification. This move was not supported and the Swedish astronomer Knut Lundmark soon wrote to W. Campbell, the Lick Observatory director: 'I admire Mr. Bigourdan's work but I must repeat that I felt disappointed at reading his report'.

In 1922, a well-known American astronomer, Vesto Slipher, was elected as president of the commission. He began to concentrate ideas on the study of nebulae.

Hubble was at that time greatly influenced by Jeans' concepts of galactic evolution. However, he was not alone in thinking that the diversity of shapes of extragalactic nebulae implied an evolutionary meaning. One of the documents of the American division of the committee on nebulae stated: 'We seem to be succeeding with the evolutional sequence classification of the stars, and we may look forward with some hope to a time when something of the sort can be attempted with the nebulae'. Hubble noted in one of his letters to Slipher: 'The observer may well look to Jeans' theory for the thread of physical significance that shall vitalize a system of classification of nongalactic nebulae'. Still, he was extremely careful. 'In the scheme presently to be proposed, a conscious attempt was made to ignore the theory and arrange the data purely from an observational point of view'. This was the way Hubble introduced his new classification scheme, in its practically final form that he sent to Slipher on 24 July 1923:

I am sending you some notes on a system of nebular classification based upon photographic images. Instead of working them up into an article for publication I thought it would be better to offer them to the committee on nebulae as a basis for discussion, the outcome of which might be a system of classification approved by the committee and sanctioned by the I.A.U.

Hubble's classification was elegant and simple. He separated all extra-galactic nebulae into three groups: elliptics E, spirals S and irregulars I. Nebulae of the first group have elliptic shape, from distinctly prolate to distinctly oblate. Those of the second group, with characteristic spiral arms, are classified into normal spirals S, in which the arms issue from the central region, and barred spirals SB with a linear bar going through their centre and from which the arms issue. The nebulae are denoted by the letters a, b and c, depending upon how the arms are twisted, how rapidly they unwind away from the centre, and are referred to as early, intermediate and late nebulae. Finally, the last type of nebulae is the group of irregular nebulae.

The description of this classification was sent to the members of the committee on nebulae. Hubble again wrote to Slipher in February 1924: 'Mr. Hale thinks I should publish the system of classification. I would prefer that it go through the committee if that is feasible within a reasonably short time'.

The second General Assembly of the Astronomical Union was convened in summer 1925 in Cambridge, England. The committee on nebulae gave a cold shoulder to Hubble's new proposals. Its members were of the opinion that the classification again operated with notions that appealed to certain poorly investigated physical properties of nebulae. They preferred a purely descriptive system, and refused to choose Hubble's classification as the official one.

In spring 1926, the widow of the well-known British astronomer Isaac Roberts, Mrs Dorothy Klumpke-Roberts (she was on the committee) published a report on the work of this committee in a French science-popularising journal. She characterised Hubble's classification as excellent and gave its first printed description. Now the classification became available not only to members of the committee but also to all astronomers.

In September 1926, Hubble sent to the *Astrophysical Journal* a large paper on extragalactic nebulae, probably because he was now sure that the committee would not accept his classification. The issue in which the paper appeared was published in December of the same year. A brief outline of the classification was printed in the August issue of *Publications of the Astronomical Society of the Pacific*. Hubble not only gave descriptions but also showed photographs of all types of extragalactic nebulae. About 97% of all nebulae possessed central symmetry, and only 3% were classified as irregular. Summarising the results of his work, Hubble wrote:

Although deliberate effort was made to find a descriptive classification which should be entirely independent of theoretical consideration, the results are almost identical with the path of development derived by Jeans from purely theoretical investigations. The agreement is very suggestive in view of the wide field covered by the data, and Jeans' theory might have been used both to interpret the observations and to guide research. It should be borne in mind, however, that the basis of the classification is descriptive and entirely independent of any theory.

Having classified nebulae with known apparent integral magnitudes, Hubble started their statistical analysis. He discovered that these magnitudes and the diameters of nebulae are interrelated, and the parameter of this relation changed in a regular manner from circular to more and more oblate elliptic nebulae, and also from early to late spirals. At the time Hubble obtained this result, it appeared especially interesting. From today's standpoint, however, another aspect was much more important. Hubble already knew distances to seven nebulae and was able to evaluate their total true luminosities; in astronomer's lingo, he knew their absolute stellar magnitudes. The absolute magnitudes of the brightest stars in these nebulae were also measured.

It was found that the absolute magnitudes of the nebulae were very similar. If this conclusion was correct, a way was open to deduce from the apparent stellar magnitudes, at least approximately, the distances to the nebulae in which no individual stars could be resolved even by the most powerful telescopes. Even today, apparent magnitude remains the only measure of distance to a galaxy when all other methods have failed.

Hubble also obtained another very important result. He discovered, from the counts of nebulae in a given range of magnitudes, that the fainter the galaxies, the greater the number of them found on the plates. The dependence was as one would expect for a sufficiently uniform distribution of objects in space. On average, roughly one nebula was present per 10^{17} cubic parsecs; that is, they were separated by distances of about 570 kiloparsec.

In 1923, Hubble submitted the classification that he had developed to the members of the committee on nebulae of the I.A.U, and three years later it was published in the *Astrophysical Journal*, the best-known astronomical journal. Suddenly, in spring 1926, Knut Lundmark proposed his system of classifying nebulae, which greatly resembled Hubble's proposal.

Lundmark had also divided all nebulae into galactic ones and those he called 'anagalactic', and grouped the latter into elliptics, spirals and objects of the type of the Magellanic Clouds, that is, into irregular galaxies.

On 22 June, Hubble wrote an angry letter to Slipher:

I see that Lundmark has published a 'Preliminary classification of nebulae' which is practically identical with my own, except for the nomenclature. He calmly ignored my existence and claims it as his own exclusive idea. I am calling this to your official attention because I do not propose to let him borrow the results of hard labor in this casual manner.

Hubble also sent a sharp letter to Lundmark. 'Can you suppose that colleagues will welcome your presence when they realize that it is necessary to publish before they discuss their work?' Hubble never got over his dislike of Lundmark. Writing to Shapley on a different matter, Hubble mentioned his distrust of Lundmark as scientist and person:

Your findings ... emphasize the need for independent research and judgments concerning whatever he publishes. Here, as elsewhere, he has mixed the good with the bad, facts with fancies, in such a manner that the general significance of his results is extremely misleading.

There was definitely a basis for such letters. Lundmark worked beside Hubble at Mount Wilson for more than two years, after his arrival in June 1921. There seemed to be nothing to spoil relations between the two scientists, and they were studying together an interesting supernova, Z Centauri. This was the time when Hubble developed his first version of the classification of nebulae and published it in the *Astrophysical Journal*; Lundmark simply could not be unaware of it, even if he lived at this time not in the USA but in his native land, Sweden. It was also possible for him to read about Hubble's work in another journal, *Publications of the Astronomical Society of the Pacific*, where W. S. Adams described the successes of the observatory during the 1922–3 period.

There is another aspect that is difficult to explain away in this unpleasant collision. Lundmark was present at the meeting of the committee on nebulae when Hubble's proposal was discussed. However, his paper does not even hint of that which Lundmark could not help hearing at Cambridge. He mentioned Hubble just once, only in connection with Hubble suggesting the term 'galactic nebulae'. This all looks very strange.

The American researchers R. Hart and R. Berendsen tried to understand how the galaxy classification was born. 'The validity of Hubble's claim has not yet been solved, although two observations can be made. First, Lundmark's paper, which apparently was not based on the work of Jeans as was Hubble's, contained no appeal to an evolutionary scheme; and second, Lundmark has been working on classification at least as early as 1922'.

We can hardly accept this interpretation. There is no doubt that Hubble liked Jeans' theory, but he stressed the independence of his scheme from that of Jeans on many occasions. In May 1922, Lundmark wrote to Campbell that one of the directions of his work at Mount Wilson, which had interested Hale, would be a 'statistical investigation of known spirals in connection with the question of classifying the non-galactic nebulae'. These words hardly imply that Lundmark was working on this problem simultaneously with Hubble and independently of him. Historians of astronomy have been unable to find other, more solid evidence of this.

Only one thing supports Lundmark's position. This is the fact that his scheme cannot be said simply to repeat Hubble's classification. For instance, Lundmark classified elliptical nebulae not by the apparent flatness of the object but by the concentration of light towards the centre of the nebula. The criteria for spiral nebulae are also more varied: the concentration of light, the general shape of spiral arms, and so forth.

In his large and interesting paper of 1927 on extragalactic nebulae, Lundmark warmly acknowledged the help of quite a few Mount Wilson colleagues but never mentioned Hubble. He responded to Hubble's accusations with equal severity. Yes, indeed, he had been at Cambridge but not yet in the capacity of a member of the commission on nebulae, and he knew nothing about Hubble's memorandum, nor about his work on classification after 1922. Lundmark's own scheme does not at all repeat Hubble's work. Besides, if the past is to be reconstructed at all, Lundmark sarcastically remarked, the terms 'elliptical' and 'spiral' nebulae were first used by S. Alexander and F. Ross long before Hubble, in the middle of the last century.

Many years have gone by since that time, and now only historians of science may be interested in the quarrel between the two astronomers. As for science, it has made an unequivocal choice in favour of Hubble's classification, although it has never been adopted as official. The outstanding astronomer Walter Baade said:

It is a very simple one but... There is really not much sense in making a system that covers all the little details of spiral structure. About the merits of Hubble's system I can speak from experience; I have used it for 30 years, and, although I have searched obstinately for systems that do not fit it, the number of such systems that I finally found – systems that really present difficulties – is so small that I can count them on the fingers of my hand... If you eliminate the double systems, I am sure that the number of exceptions is unbelievably small, so efficient is the system.

However, Hubble's system was not accepted overnight. In 1927, Reynolds voiced his objections. He thought that the classification was too simplified, disregarding many details in nebulae. Reynolds wanted astronomers to return to his proposal, with new criteria added. In his response, Hubble reminded Reynolds of his guiding principle: 'A great range in structural details is admitted, and for this reason a first general classification should be as simple as possible'. Shapley also made an attempt to develop a new compromise system, by combining Lundmark's and Hubble's approaches. However, nether Reynolds', nor Lundmark's, nor Shapley's classifications survived.

Hubble's classification has served astronomy for more than sixty years now, and its essential features have not been affected by modernisations.

In 1979, K. Lang and O. Gingerich published a very interesting collection, *A Source Book in Astronomy and Astrophysics, 1900–1975*, in which they reproduced, word for word, all the important publications which determined the face of our science over three quarters of a century. Very high honour was assigned to three of Hubble's papers: the first of them was his paper on the classification of extragalactic nebulae, the second paper proved that such nebulae are huge stellar systems far away in space beyond our galaxy, and the third paper formulated the law of their motion, Hubble's Law.

The island universe

Astronomers had been studying numerous nebulae for over 150 years since William Herschel's time, measuring their coordinates and describing their characteristics. Nevertheless, the true nature of nebulae remained a mystery. Some believed that nebulae were connected with the world of stars surrounding us, others dared to conjecture that they are independent stellar systems similar to our galaxy but removed to tremendous distances.

New adepts were attracted sometimes to one, sometimes to the other of the competing points of view.

On 26 April 1920 the National Academy of Sciences of the USA had organised a debate between two leading astronomers, Harlow Shapley and Herbert Curtis. Although the main topic of the debate was the structure of our galaxy, the nature of spiral nebulae was also discussed.

Curtis stated his conviction:

The evidence at present available points strongly to the conclusion that the spirals are individual galaxies, or island universes, comparable with our own galaxy in dimension and in number of component units.

To which Shapley replied:

It seems to me that the evidence... is opposed to the view that the spirals are galaxies of stars comparable with our own. In fact, there appears as yet no reason for modifying the tentative hypothesis that the spirals are not composed of typical stars at all, but are truly nebulous objects.

This dispute did not discover the truth, nor could it do so at the time: the opponents lacked decisive arguments and, especially, data on distances to nebulae.

A very large nebula in Andromeda, known to mankind for at least a thousand years since it had been mentioned in a treatise by Al-Sufi, the outstanding Arabic astronomer of the 10th century, stands out among all other nebulae. The history of the study of the Andromeda nebula in Europe began in 1612 when the German astronomer Simon Marius discovered it through a telescope; Marius knew nothing about Al-Sufi's book. Later J. Flamsteed, G. Cassini, C. Messier and many others observed it through their inevitably imperfect instruments (Messier placed it in his famous catalogue as No. 31). William Herschel observed it on many occasions. His notion of the Andromeda nebula changed continually. Sometimes it seemed to him that the nebula was about to be resolved into individual stars, and sometimes he thought of it as a diffuse nebula, resembling many others which were later proved to be gaseous clouds.

In 1885, a new bright source flared up in this nebula: the famous supernova S Andromedae. Several years later, a rich British amateur astronomer, Isaac Roberts, obtained a series of photographs of the nebula through a telescope with a 20-inch mirror. His excellent photographs of the nebula revealed spiral arms in which it was already possible to identify stars as individual dots. Spectral observations also gave evidence that the object was not a gaseous blob. William Huggins, another English amateur astronomer, was unable to see in his visual spectroscopes those bright emission lines in the spectrum of the nebula's core which he could expect to find by analogy with many other diffuse objects in the Milky Way. In the last year of the 19th century, C. Scheiner in Potsdam was able to photograph the spectrum of the nebula. No emission lines were apparent and Scheiner had to conclude that the Andromeda nebula (in fact, he was speaking about its central, bright part) was a stellar system.

The decisive factor in clarifying the problem of the nature of the Andromeda nebula and other similar objects would be the measurement of distances to them. Julius Franz in Königsberg made an unsuccessful attempt at measuring the trigonometric parallax of the supernova of 1885. The Swedish astronomer Karl Bolin also tried to measure the parallax of the nebula on photographs. According to his results, the nebula was at a distance of only 19 light years from us – this is a fantastically erroneous result. Other astronomers also failed to obtain any reliable data on this distance.

Progress became possible only when new stars were discovered in the Andromeda nebula. While scanning photographs of one of the spiral nebulae, the Mount Wilson astronomer George Ritchey suddenly noticed that a new star (a nova) had appeared in it. We know now that this was not a nova in the usual sense of the word, but rather a supernova. At that time, however, astronomers did not realise that such events of stars suddenly flaring up have to be separated into two different classes. It was decided immediately to analyse photographs of other nebulae as well. Soon two novae were discovered on the old plates of the Andromeda nebula. Ritchey's communication published in 1917 was the first information on novae in this nebula. The discoveries of other novae were reported in the same and later years by Shapley, Ritchey, Duncan, Sanford and Humason. Discoveries followed one after the other, and the 21st nova in the Andromeda nebula was discovered in 1922.

Curtis was the first to realise that the bright star of 1885 was drastically different from other, weaker novae in Andromeda, and thus should not be taken into account when determining their distances. Using only ordinary novae and comparing them with similar objects in the galaxy, Lundmark found in 1919 that the distance to the Andromeda nebula was 550 000 light years, or 170 000 parsec. This was a very significant step forward.

Nevertheless, one could still not be sure that such measurements put the novae in the Andromeda nebula in correspondence with novae in our galaxy. The results of indirect methods could not be regarded as very reliable either. One needed objects that could be used as reliable indicators of distance.

In the summer of 1923, Hubble began vigorously to observe the nebula using the 60- and 100-inch reflectors, mostly in order to accumulate enough data for a statistical investigation of novae. Already the first good plate exposed on 4 October using the 100-inch instrument revealed two novae and one faint variable star. This variable was his major discovery. In the catalogue of plates that survived in the archive, listing the brightness of the variable, Hubble wrote: 'Found on this plate. Oct.10/23'; on the glass of the plate, Hubble crossed out the letter N marking the star (a nova) and put in in large letters 'Var!' (variable). Hubble found this star on several dozens of negatives, exposed as early as autumn 1909 when Ritchey was working on the 60-inch telescope. He was able to measure the period of the variable on 23 October and plotted its brightness curve. The plates were too scattered in time to satisfy him, so Hubble also wanted to acquire a continuous sequence of observations. The nasty weather of November and December of 1923 was over and January of the new year had begun. Unexpectedly, stable clear weather set in in February. Hubble photographed the Andromeda nebula for nearly a week from 2 to 7 February. The star's brightness was increasing very rapidly. There was no doubt that this was a typical Cepheid caught at the rising part of its light curve.

No one astronomer before Hubble had ever tried to find Cepheids in the Andromeda nebula. The collection of photographs piled up by Shapley was lying idle. Only in September 1924 did Lundmark, who knew nothing about Hubble's success, mention at a meeting of the German astronomical society that it was necessary to look for Cepheids in the Andromeda nebula, in order to determine reliably the distance to it.

It would be difficult to find objects in astronomy that played a more important role than Cepheids did. Miss Henriette Leavitt at the Harvard Observatory had established as early as 1908 that brightness variations of variable stars in the Small Magellanic Cloud had variation periods that were related to their brightness, that is, to their apparent stellar magnitudes. The stars were virtually at the same distance from the Earth. Miss Leavitt concluded that 'their periods are apparently associated with their actual emission of light'. The behaviour of these stars resembled that of variables known in globular clusters. The first to realise that these variables were Cepheids was the famous Danish astronomer Einar Hertzsprung. He also made an attempt to find a relation between the period of Cepheids and their true emission of light, that is, to find the period–luminosity relationship.

Astronomers now had in their hands a powerful method of measuring distances. In principle, it is sufficient to find the period of brightness variation of a Cepheid (this is not very difficult to do) and then determine the absolute luminosity of the star using the period–luminosity curve. Next, a comparison of the apparent and the true magnitudes gives an estimate of the distance to the Cepheid; if this Cepheid belongs to, say, the Andromeda nebula, then one knows the distance to the nebula itself.

Hubble first disclosed his results on 19 February in a letter to Shapley, who was an expert on variables. He wrote:

You will be interested to hear that I have found a Cepheid variable in the Andromeda nebula (M31). I have followed the nebula this season as closely as the weather permitted and in the last five months have netted nine novae and two variables.... The two variables were found last week ['confirmed' would be a better word here, rather than 'found' – A.S.]. No. 1 is roughly 16′ preceding the nucleus, but well within the borders of the arms, and is situated on a background of faint mottled nebulosity. Magnitudes were estimated rather hastily from a set of comparison stars and a light curve was constructed covering all available observations from 1909 down to date. The zero point in the comparison scale was extrapolated from your pg magnitudes, making what seemed to be a fair allowance for distance correction. I believe the range of the variable cannot be as much as 0.3 in error, nor the median magnitude more than 0.5.

Enclosed is a copy of the normal light curve, which, rough as it is, shows the Cepheid characteristics in an unmistakable fashion. Am I right in supposing it a typical cluster type curve? The period of 31.415 days corresponds to $M = -5.0$ on your period–luminosity curve. The medium sg magnitude of about 18.5 needs an uncertain correction for color index. Seares suggests 0.9 as a maximum, although your period–color curve for the Magellanic cloud calls for a higher value. With Seares' value, the median pv magnitude is 17.6 and the distance comes out something over 300 000 parsecs. If the stars were dimmed materially by shining through nebulosity, the distance would be correspondingly reduced... (see Figure 1).

The most important result is given in the penultimate sentence of this quotation. The Cepheid made it possible to establish reliably that the Andromeda nebula to which the star definitely belonged lay at a distance of about a million light years from us. This immediately implied that the Andromeda nebula is located far beyond our stellar system and both it and our galaxy, and most likely weaker nebulae as well, were more or less similar islands of the universe. That was the end of the concepts which had been Shapley's life. Shapley immediately understood that. Payne-Gaposchkin recalled: 'I was in his office when Hubble's letter came, and

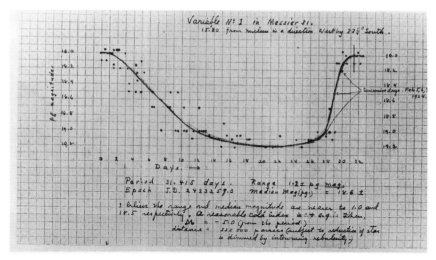

Figure 1. The light curve of the first Cepheid discovered
by Hubble in the Andromeda nebula. The figure is taken
from Hubble's letter to Shapley

he held it out to me: "Here is the letter that has destroyed my universe",
he said'.

Shapley wrote back to Hubble on 27 February:

Your letter telling of the... variable stars in the direction of the Andromeda
nebula is the most entertaining piece of literature I have seen for a long time...

Hubble also discovered other faint variables in the spiral arms, but was
not yet able to study them. All this was described in detail in the
annual report of the observatory, although the most important result –
an estimate of the distance to the Andromeda nebula – was not even
mentioned. Presumably, W. Adams, who replaced Hale as the director
at Mount Wilson, chose to be cautious and to wait for the results of
studying other stars.

In fact, the observational season 1923–4 was very successful for
Hubble. He discovered ten novae and thus enriched the list of such
objects in the Andromeda nebula to 32. Some of these stars proved to
be very intriguing. For instance, one nova varied very slowly and left its
trace on plates from October 1921 and the next five and a half years.
Even today, we do not know any other nova in this nebula that would
behave in this manner. Three more novae were found at such large
distances from the centre that they had not previously been observed.

Hubble had a fortunate ability to work simultaneously and vigorously in several fields. In these years he was studying, without leaving the Andromeda nebula, two more stellar systems, NGC 6822 and the Triangulum nebula, M33.

The first of them was discovered in 1884 by the American astronomer Barnard at his 5-inch refractor. This was a faint, hardly discernible object. True, Barnard thought that this object became brighter when he observed it the following year through a different instrument; he even classified it as a variable nebula. Max Wolf studied it on photographs exposed at Heidelberg in 1906–7 and described the object as a group of small nebulae. Its nature became clear only in 1922. Perrine in Cordoba photographed this sufficiently southern object with a 30-inch reflector. It looked very much like a miniature Magellanic Cloud and consisted of stars and several diffuse nebulae. It had the same appearance on the plates obtained by Duncan in July 1921 using the 100-inch telescope. Photographs made through the 10-inch camera confirmed the peculiarity of the object, and it was included in the programme for detailed study on large instruments.

In 1923 and 1924, from June to November, Hubble obtained about 40 good plates of NGC 6822 and discovered variable stars in it; some of them were Cepheids, as in the Andromeda nebula.

Later Hubble focused his attention on another interesting and large object – the Triangulum nebula M33.

The history of its study began fairly long ago. On 25 August 1764, the famous comet-hunter, Parisian astronomer Charles Messier, had discovered a new nebula, which he later included in his catalogue as number 33 (M33). The nebula attracted many observers. Using his telescopes, which were huge for his time, William Herschel observed it on many occasions. He sometimes even thought that the nebula separated into individual stars. In the middle of the last century, Lord Ross observed the nebula through a telescope with a 6-foot mirror, the 'Parsonstown Leviathan', and discovered a spiral structure made of five arms. Astronomers measured the position of a nebula and of its parts relative to the stars, in the hope that its proper motion would be measured in the future: they had no idea at the time about the distance to such objects.

The first photographs of the nebula were obtained by Isaac Roberts in the 1890s. The photograph presented in his famous 'Atlas' does not differ in any way from those obtained today. The whole nebula was a swarm of faint and slightly nebulous stars.

Later, the centre of investigations of the Triangulum nebula shifted, as had happened with the Andromeda nebula, from Europe to America, where large telescopes were being built. In 1899, it was photographed by J. Keeler on the Crossley 36-inch reflector of the Lick Observatory, and ten years later, Ritchey obtained excellent photographs on the still larger, 60-inch instrument at Mount Wilson. These photographs clearly showed individual stars; nevertheless, Ritchey refused to classify them as stars. Ritchey, who counted about two and a half thousand such objects in M33, wrote: 'All of these spirals contain great numbers of soft star-like condensations which I shall call nebulous stars. They are possibly stars in process of formation'. The concluding phrase clearly reflects the cosmogonic notions prevalent at that period.

Beginning in 1919, the Triangulum nebula was regularly photographed by Duncan, Pease and other astronomers using the 100-inch reflector. Lundmark in 1921 was perhaps the first to conjecture that Ritchey's nebulous stars are ordinary stars. Furthermore, if the brightest of them had the same absolute magnitude as their analogues in our galaxy, the distance to the Triangulum nebula should be very large: up to 300 000 parsec.

In 1922, J. Duncan discovered that the brightness of three stars in the nebula varies. The variability of one of them was discovered independently by Max Wolf.

In August 1924, Hubble included this nebula in his programme of regular observations, and received in the same season about 200 plates of the Andromeda and Triangulum nebulae and NGC 6822. Hubble was able to identify Cepheids in the Triangulum nebula, as well as in the other two systems under study.

In these months, Hubble was engrossed not only in happy and successful work but also by overwhelming personal happiness. At the end of February 1924, he married Grace Burke Leib. In all likelihood, Grace and Edwin first met in the expedition to observe the solar eclipse on 10 September 1923, where she went with her uncle, the Lick astronomer Fred Wright. Sandage thought that they met in San Diego. The Mount Wilson Observatory had also mounted a series of experiments, and the young people perhaps met during the preparatory stage, when the Mount Wilson and Lick people were working together.

The beautiful, very lively dark-eyed young woman, who looked especially frail in comparison with her tall husband-to-be, had probably taken Edwin's imagination immediately. He was 34 years old and the course of his life had been such that perhaps he had never before had any really passionate encounters. Mrs Helen Lane, Hubble's sister, remembered:

His only serious interest that I knew of was for a very lovely girl he met in Springfield, Mo, where we visited often... However, he and Elizabeth really cared for each other... but after some time, she realized that she could never substitute Mars, the nebulae and such. I really believe it was a blow to his ego, but we don't know of any other serious affair he ever had before his marriage.

Mrs Hubble recalled that she first heard about Edwin from her uncle. Her uncle was considerably older than Edwin and once said, probably as a man with much greater experience of life: 'He is a hard worker. He wants to find out about the universe; that shows how young he is'. We know this story only in Mrs Hubble's words, so that it is difficult to know what exact meaning Mr Wright put into the words about finding out about the universe. However, if we think about the general picture of the world – the nature of galaxies, their distribution in space and their striking proper motions, all this became known mostly through Hubble's work. The first steps in this direction were made at the very beginning of their happy family life.

The news of Hubble's discoveries gradually spread among America's astronomers. The public was first able to read about them in a short article, of only 30 lines in the *New York Times* of 23 November 1924. As a rule, a newspaper is a poor source of scientific information: even now newspapers abound in confusing mistakes and sometimes blunder fantastically. This time, however, the readers were in luck. An unknown reporter was mistaken only in Hubble's name but quite correct as far as the message was concerned.

Since that time, the article has never been reprinted in English or in other languages, so it is worthwhile to repeat it now, several decades later, as the first news of an illustrious result. The title of the article was: 'Finds Spiral Nebulae Are Stellar Systems. Doctor Hubbell confirms view that they are "Island Universes" similar to our own'. The article read:

Washington, Nov. 22. Confirmation of the view that spiral nebulae, which appear in the heavens as whirling clouds, are in reality distant stellar systems, or 'island universes', has been obtained by Dr. Edwin Hubbell of the Carnegie Institution's Mount Wilson Observatory, through investigations carried out with the observatory's powerful telescopes. The number of spiral nebulae, the observatory officials have reported to the institution, is very great, amounting to hundreds of thousands, and their apparent sizes range from small objects, almost star-like in character, to the great nebula in Andromeda, which stretches across an angle of some 3 degrees in the heavens, about six times the diameter of the full moon.

The reporter then said:

The investigations of Dr. Hubbell were made photographically with the 60-inch and 100-inch reflectors of the Mount Wilson Observatory, the extreme faintness of the star under examination making necessary the use of these great telescopes. The resolving power of these instruments breaks up the outer portions of the swarms of stars, which may be studied individually and compared with those in our own system. From an investigation of the photographs thirty-six variable stars of the type referred to as Cepheid variables, were discovered in the two spirals, Andromeda and No. 33 of Messier's great catalogue of nebulae. The study of the periods of these stars and the application of the relationship between length of period and intrinsic brightness at once provided the means of determining the distances of these objects.

The results are striking in their confirmation of the view that these spiral nebulae are distant stellar systems. They are found to be about ten times as far away as the Small Magellanic Cloud, or at a distance of the order of 1 000 000 light years. This means that light travelling at the rate of 186 000 miles a second (300 000 km/s) has required a million years to reach us from these nebulae and that we are observing them from the Earth by light which left them in the Pliocene Age.

With a knowledge of the distances of these nebulae we find for their diameters 45 000 light years for the Andromeda and 15 000 light years for Messier 33. These quantities, as well as the masses and densities of the systems are quite comparable with the corresponding values for our local system of stars.

Christmas in America is celebrated on a grand scale, while New Year's day is typically a modest occasion; correspondingly, the next meeting of the American astronomical society was slated for 30 December 1924 in Washington. The city met the participants of the meeting in an unwintery fashion. It was warm, people were dressed in suits or light overcoats. Winter imposed itself only on the day the participants were leaving, when a severe snowstorm almost paralysed all traffic.

The congress differed from the preceding ones only in that this time it was conducted together with the American Association for the Advancement of Science. The audience was formed of astronomers, physicists and mathematicians. Among the astronomers, quite a few were already well known or famous, or became known and famous later: Curtis, Shapley, Slipher and others. Hubble's science advisor at Chicago University, Professor Frost, came too.

About forty papers were read to the congress; when we read the list now, we realise that most of them dealt only with details. Many participants were worried about observing the total solar eclipse on 24 January, and Captain Pollock, the director of the Naval Observatory in Washington (the only federally administrated observatory in the country at that time), announced the good news that accurate time signals were to be transmitted specially, before and after the total eclipse phase.

The participants were anticipating a talk by the famous English astronomer Sir Arthur Eddington about stellar evolution. He arrived in Canada and the USA that summer, gave a number of lectures, visited the Mount Wilson Observatory, and met members of its staff, including Hubble. However, a cable came at the beginning of November with the news of his mother's death, so Eddington cut short his visit to the North American continent. Eddington's talk was replaced with a short communication by Henry Norris Russell. Russell did not manage to be present at the opening of the meeting and asked J. Stebbins during their late lunch at the hotel whether Hubble had sent in his contribution.

Russell, the director of the Princeton University Observatory, spent a long time every year at the Mount Wilson Observatory. His field of interest was exceptionally broad and he was able to recognise the outstanding importance of Hubble's achievement immediately. In fact, Russell had already written to Frank Schlesinger at the end of October that he considered Hubble a candidate for membership of the National Academy of Sciences of the USA, immediately after Hubble had published the results of his research. Indeed, the election did take place in summer 1927, after his papers on NGC 6822, M33 and M31 had been published. Russell was in close contact with various journals; in the list of the main successes of astronomy written for the editor of *Science Service*, Russell stated that Hubble's discovery was 'undoubtedly among the most notable scientific advances of the year'. He understood that Hubble's study presented to the meeting would definitely win him the prize of the American Association for the Advancement of Science, established by an anonymous wealthy member.

However, there was no contribution from Hubble. Russell remarked: 'Well, he is an ass! With a perfectly good thousand dollars available he refuses to take it'. After some discussion, Russell and Stebbins decided to send a cable to Mount Wilson to persuade Hubble to communicate immediately the main results, so that they could write there, in Washington, something like Hubble's talk. They wrote the text of the cable. Russell and Stebbins approached the hotel counter in order to write a telegram. Turning to go to the telegraph, Russell suddenly noticed a large package addressed to himself, and Stebbins immediately noticed Hubble's name in the upper left-hand corner of the package. This was the paper for which they had waited so long.

Russell read Hubble's paper to the joint morning session of astronomers, physicists and mathematicians at the George Washington University on 1 January 1925; the paper title was 'Cepheid Variables in Spiral Nebulae'.

The paper stated that the only indication of the existence of stars in the Andromeda nebula was, until recently, the presence of novae, and in the Triangulum nebula it was the presence of three variable stars discovered by Duncan. However, the outer parts of the two objects, when photographed through large telescopes, evidently separated into numerous stellar images. By blink-comparing the plates, of which Hubble accumulated about 200, he had identified a considerable number of variables. After the newspaper publication, Hubble continued to work hard and, by the end of 1924, had already found 36 variables and 46 novae in M31, including those 22 which were found before Hubble by Mount Wilson observers. Together with Duncan's objects, M33 evinced 47 variable stars. Hubble had also discovered the first nova, one of the very few ever to flare up in this galaxy. Light curves were plotted for the 22 stars of M33 which had proved to be Cepheid variables. Twelve Cepheids were also studied in M31. Quite a few variables that were not covered by this study most probably belonged to the same type.

The Cepheid variables of the two systems satisfied the 'apparent magnitude versus logarithm of period' relationship. Its comparison with the 'absolute magnitude versus logarithm of period' relation established by Shapley in 1918 made it possible to evaluate that the distance to the M31 and M33 nebulae is the same, equal to 285 000 parsec.

Hubble came to a clear-cut formulation of his three basic hypotheses: first – the Cepheids indeed belong to spirals, second – there is no appreciable light absorption in spirals which would diminish their brightness and, finally, third (most important) – the nature of the variability of Cepheids is identical throughout the universe. This last principle, extended to all other objects with identical characteristics, is the basic one for measuring distances in the world of galaxies.

M31 and M33 were no exception. Variables were also found in M81, M101 and NGC 2403, but so far there was an insufficient number of plates to perform a careful study.

After the talk, it was clear to everyone that this was the main event of the congress.

The council of the society selected Hubble's work as undoubtedly deserving the prize, and entrusted it to Russell and Stebbins to watch over its submission in the proper manner to the committee on prizes. Russell prepared all the necessary papers. On the same day, Stebbins, as secretary to the American Astronomical Society, sent to the committee an enthusiastic letter. Having characterised Hubble's communication to the congress, Stebbins ended his letter with the following words: 'This paper is the product of a young man of conspicuous and recognized ability in a field which he has made peculiarly his own. It opens up depths of space previously inaccessible to investigation and gives promise of still greater advances in the near future. Meanwhile, it has already expanded 100-fold the known volume of the material Universe and has apparently settled the long-mooted question of the nature of the spirals, showing them to be gigantic agglomerations of stars almost comparable in extent with our own Galaxy'.

Hubble's report had already appeared in the American journal *Popular Astronomy* in April of that year, and was soon reprinted by the British journal *Observatory*; soon it was known to the entire world community of astronomers. (Curiously, it appeared in the *Publications of the Astronomical Society* itself only in 1927, when the proceedings of several meetings were jointly published.)

Russell, who wrote a regular column in *Scientific American*, described Hubble's discovery in the March issue. In Russia, the astronomical community was able to read about this news in the review written by V. Maltsev for the August issue of the journal *Mirovedeniye*. Hubble's name was unlucky again: in Russian it was for a long time written in various ways, all of them wrong.

On 13 February, *Science* published a brief notice that the committee had decided to share the prize between two scientists, Dr Edwin Hubble and Dr Cleaveland, specialist on termites, and a month later gave a more detailed description of the contributions of the prize winners. The *Publications of the Astronomical Society of the Pacific* informed its readers about Hubble's prize. Stebbins immediately congratulated Hubble, remarking that he was sorry that Hubble had not recieved the entire prize. The prize meant public recognition of Hubble's standing in science; his name appeared for the first time in *Who Is Who in America* for 1924–5.

At that time, 500 dollars was a substantial sum. As a rule, the Association for the Advancement of Science allotted prizes from 100 to 300 dollars. If the Hubbles decided to move from the hot Californian climate to some colder area, Hubble would be able to make his young wife a gift of, say, a leopard fur coat which had been advertised on the same page of the newspaper where his discovery had first been written up. And he would have enough money left for tobacco for his cherished pipe.

From what we know, Hubble did not really expect the prize. He was very glad, and on 19 February he wrote to Russell:

Dear Mr Russell,

The award came as a joyous surprise. I had supposed that only finished work could be considered. We realize that the business was about 99 per cent Russell and 1 per cent Hubble. It is impossible to express my thanks in any adequate manner, but rest assured that I am tremendously appreciative of your suggestion (very pointed) to send in a paper, and of your very good offices in urging its consideration on the various committees.

The real reason for my reluctance in hurrying to press was, as you may have guessed, the flat contradiction to van Maanen's rotations. The problem of reconciling the two sets of data has a certain fascination, but in spite of this I believe that the measured rotations must be abandoned. I've been examining the measures for the first time and the indications point steadily to a magnitude error as a plausible explanation. Rotation appears to be a forced interpretation, especially in the cases where the measured total displacements are large – M81, M33, and M51, and really the only strong argument in its favor is the rather staggering agreement in direction of motion with the direction of the arms. I am anxious to show you the evidence when you arrive.

Meanwhile a mass of undigested data is accumulating from the observations – star counts and color plate for M33, novae in M31 (6 on one plate), variable in other spirals, evidence of resolution in irregular non-galactic nebulae, etc.

The really big advance as I see it, is the possibilities of applying the usual methods of stellar investigations to the spirals.

Sincerely.

Edwin Hubble.

Mrs Hubble is waiting to thank you properly for your splendid efforts in our behalf.

From the standpoint of the history of science, the significant message is not so much in the words of thanks as in the motives behind the almost year-long delay in the announcement of the discovery.

The Dutch astronomer Adrian van Maanen started working at the Mount Wilson Observatory in 1912. He first worked in the solar research group, and two years later he began his individual research, measuring the exact positions of stars and parallaxes. Then he got interested in spiral nebulae and in 1921 published first a preliminary study and in 1923 a complete study of the rotation of M33. The plates he used for this purpose were separated by an interval of twelve years. Having measured the positions of numerous details in the spiral arms of the nebula with respect to the surrounding stars on each plate, van Maanen found the proper motion of the object as a whole, and having eliminated this motion, decomposed the remaining motion into two components: the radial and the rotational motions. He found that the latter component was quite appreciable. This meant that the nebula was rotating and, by van Maanen's estimate, made one revolution in 60 000 to 240 000 years, depending on the distance from the centre.

Van Maanen's results seemed very conclusive to a number of astronomers, especially because he found similar rotation in other spiral nebulae. All this looked like a clear illustration of motion in the protoplanetary cloud of Jeans' theory. Van Maanen was very sure of his conclusions at that period. However, the discovery of Cepheids in the Triangulum and Andromeda nebulae made astronomers wary.

If Hubble was right in his evaluation of distances, the peculiar rotational motions, recalculated into linear velocity units, exceeded the speed of light!

The best judge of the results on peculiar motions is the time: the longer the interval between the compared plates, the more accurate the results. In 1935, Hubble made good his promise to Russell, being at last able to offer the final proof that van Maanen's conclusions were spurious.

After a great number of years, things do not look as dramatic as they had appeared in the past. One scientist obtained a wrong result, another criticised it, the first had to agree, and the dispassionate objective lens of history finally fixed them both in a friendly handshake, smiling to each other. However, this is not often the case. The Canadian historian of science N. Hetherington has analysed very many documents – letters and notes written by various persons – and was able to reconstruct the true story of the relationship between Hubble and van Maanen.

When S. Nicholson, van Maanen's Mount Wilson colleague who repeated his measurements on the insistence of the latter, showed Hubble the data, Hubble realised that these results fell 'within the uncertainties of the material and hence do not themselves definitely indicate rotation'. On 11 March 1925 Hubble wrote to Shapley that van Maanen rejected his critical comment 'in a wildly personal manner'. In 1968, Mrs Hubble recalled in her letter to the English historian of astronomy M. Hoskin: 'Van Maanen had refused to discuss the subject or to re-examine his work. Edwin had decided to ignore the discrepancy for the time and go on with his work. A friend of his told me years afterwards that he [Hubble] said to him "They asked me to give him [van Maanen] time. Well, I gave him time, I gave him ten years"'. Hubble was sure that he was right, but rebelling against van Maanen meant causing him loss of face in public, and might have endangered his own reputation; all the more so since Hubble had worked at the Mount Wilson Observatory for

Plate 1. Edwin Hubble with his brothers and cousins (around 1900).
Left to right: Bill Hubble, Henry Hubble, Nellie James, Edwin Hubble,
Virgil James and Cecil James.

Plate 2. Hubble with friends at the University of Chicago (1906–10).

Plate 3. Hubble with his sister Lucy during World War I (around 1918).

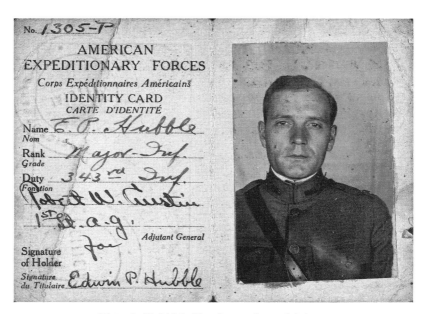

Plate 4. Hubble's identity card as a Major
in the American Expeditionary forces in Europe.

Plate 5. The Mount Wilson Observatory, showing the domes
of the 60- and 100-inch reflectors.

Plate 6. Hubble at the Mount Wilson Observatory in 1923,
during observations of the Andromeda nebula.

Plate 7. The 100-inch reflector at the Mount Wilson Observatory.

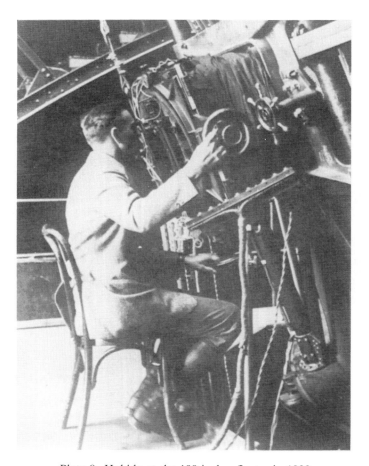

Plate 8. Hubble at the 100-inch reflector in 1923.

Plate 9. Hubble's house in San Marino, California (around 1923).

Plate 10. Mrs Grace Hubble in 1931

Plate 11. Hubble and Richard Tolman with a model of the
proposed 200-inch reflector in 1931.

Plate 12. Hubble on his 30th birthday.

Plate 13. Hubble with the English astronomer James Jeans
at the 100-inch reflector (around 1931).

Plate 14. From left to right: Walt Disney, Hubble and Julian Huxley in 1940.

Plate 15. Hubble, with Lt. Mario Cantie, witnessing a test in the supersonic wind tunnel at Aberdeen Proving Grounds on 8th May 1945.

Plate 16. Hubble fishing in Colorado.

Plate 17. Hubble at the guiding eyepiece of the 48-inch Schmidt telescope at Mount Palomar.

only four years while van Maanen had worked there three times as long. So Hubble chose to remain silent.

Time was flowing on and van Maanen's results were not exposed to serious criticism; even new evidence was presented time and again that other nebulae were found to revolve. One could hear in the Royal Astronomical Society of England that Hubble's conclusions could be accepted if it were not for van Maanen's data. That was when Hubble decided to carry out new measurements with new plates and, presumably, had to disclose his intentions. The administration of the observatory was nervous: they were afraid that a scientific debate might turn into an uncivilised quarrel. Seares, who was responsible for the Mount Wilson publications, wrote in January 1935 to Hale, who was absent at the time:

For two men in the same institution there is opportunity for personal contact and for direct examination of each other's results, and hence for the private adjustment of difference in opinion. The institution itself, it seems to me, is under obligation to see that all adjustment possible is made in advance of publication. If agreement cannot be obtained it may be necessary for the institution to specify how the results shall be presented to the public. In that event, however, there must be opportunity for the expression of individual opinion, but any such expression should be concerned only with the scientific aspects of the question at issue.

The institution has, I think, the right to enforce this procedure; but in certain cases it may be wiser to waive its technical right and say to a dissatisfied individual, 'Print what you like, but print it elsewhere'.

In other words, as Hetherington wrote, 'The Mount Wilson bureaucracy did not welcome a public dispute between two of its members'. Hubble found a way to make the new study of galactic rotation least irritating for van Maanen and most objective. The measurements involved Hubble himself and also Nicholson and Baade, invited by him; the results were also analysed by Seares. All in all, they studied four galaxies: M33, M51, M81 and M101. According to van Maanen's data, they could expect shifts on plates exposed after twenty years to reach 15 to 20 micron. In fact, shifts were only of the order of 1 micron, that is, they were within the measurement errors. However, van Maanen repeated his measurements on M33 and M74, using plates differing in time by nine years, and again obtained a rotation effect exceeding measurement errors. In his short note to the *Astrophysical Journal*, which was printed on the next page after Hubble's communication, van Maanen made an attempt to interpret his new result as a partial confirmation of his former conclusions. Nevertheless, he had to agree that the results 'make it desirable to view the motions with reserve'. However, the observatory's report stated unambiguously:

Although the anomaly of van Maanen's results remains unexplained, the recent investigations apparently remove the one outstanding discrepancy in the field of nebulae research.

Curtis, a supporter of the island universe theory, warned as early as 1920 that the island universe theory must be definitively abandoned

should the result of the next quarter century show close agreement among different observers to the effect that the annual motion of translation or rotation of the spirals equal or exceed $0''.01$ in average value.

The problem was solved earlier. Van Maanen's results had to be rejected, and the island universe theory became the cornerstone of our understanding of the surrounding world.

Investigation of the Andromeda and Triangulum nebulae was advancing, very rich observational material had been compiled and so the time had come to summarise the results.

In 1926 and 1929 Hubble published two fundamental papers. The titles were almost identical: 'A Spiral Nebula as a Stellar System', followed by 'Messier 33' in the first and 'Messier 31' in the second of the two papers (i.e., the Andromeda and Triangulum nebulae). These were large papers, of several dozens of pages each, where Hubble presented the totality of his observational work.

He conclusively demonstrated that the so-called 'nebulous' stars of M33 are ordinary stars. Among them, he found variable stars and added 42 to the three already known. Thirty-five variables proved to be typical Cepheids, whose brightness varied with periods from 13 to 70 days. Hubble was lucky to notice the explosion of two novae. The brightest non-variable stars belonged to the classes of blue or white stars, their luminosity function (the distribution of stars over luminosity) being the same as the luminosity function of similar stars in our galaxy, in the vicinity of the Sun. The spiral arms revealed bright diffuse nebulae, and the relationship between their sizes and the brightness of stars causing their emission did not differ from those we observe in our galaxy. Hence, they also contained high-temperature stars of O and B classes. Using the latest data on the period–luminosity curve for Cepheids, Hubble evaluated the distance to the Triangulum nebula as 263 000 parsec, while its diameter was found to be 4600 parsec.

The Andromeda nebula, which Hubble placed at a distance of 275 000 parsec from us, was now estimated to be an even larger stellar system than before, 64 000 parsec in diameter.

The investigation of the Andromeda nebula was based on an enormous amount of observational data: 350 plates exposed on 60- and 100-inch Mount Wilson reflectors. Two thirds of these plates had been obtained by Hubble himself during the five previous years. The outer parts of the Andromeda nebula resolved into a multitude of stars. Only the central part still looked diffuse and its stellar nature was finally demonstrated only twenty years later, not by Hubble but by Baade. Hubble found in this stellar population the same objects that astronomers saw both in our galaxy and in the Triangulum nebula. Among the 50 variables he was able to identify, 40 were shown to be Cepheids with periods from 10 to 48 days. Only one of them varied its brightness with a period of almost six months. As a result of Hubble's search, 85 novae became known in the Andromeda nebula (including the famous 1885 supernova). The novae were concentrated towards the centre of the Andromeda nebula; each year about 30 of them occurred. At their maximum luminosity, they became as bright as the brightest non-variable stars. Their absolute luminosity was evaluated on the basis of the distance to the Andromeda nebula determined via Cepheid variables.

In the spring of 1928, Hubble went to England. He had last visited Britain four years previously, when he described the classification of galaxies to his British colleagues. On 9 March Hubble was present at a meeting of the Royal Astronomical Society and met Sir Arthur Eddington, Smart, Jeans and others. He was greeted very warmly. The president, Dr Philips, said:

We have amongst us this evening a distinguished astronomer from the Mount Wilson Observatory, Dr. Hubble. We have followed with great interest the work he has been carrying on on remote spirals. I am sure we should be delighted if he would address us.

Hubble's comprehensive work on the Andromeda nebula was not yet in print (he had submitted it to the journal in December), so everyone wanted to know the details. Hubble decided to choose the novae results as his topic. One possible reason for this choice was, perhaps, that the English astronomer Stratton was to speak at this meeting about novae; or perhaps he chose this subject because he, together with Duncan, had studied the expansion of the shell of the brightest galactic nova of our century, which occurred in Aquila in 1918. Having measured the angular expansion of the envelope and its radial velocity, Hubble and Duncan estimated the distance to the nova and its absolute magnitude. Three years later they returned to studying this star and published another paper.

Hubble mentioned the large number of novae discovered in the Andromeda nebula and talked about the frequency of novae and their apparent distribution. It was interesting that, on average, these novae were much fainter than the one in Aquila. However, Hubble did not pay much attention to this difference at the time.

'I too should like to express my admiration and to add my congratulations', was Jeans' reaction.

'I am sure we all have the highest appreciation for the wonderful work Dr Hubble has been carrying on' – such was the conclusion of the President of the Royal Astronomical Society after Hubble's talk.

The Royal Society elected Hubble to its membership. This was the first official recognition of his work outside the USA.

Hubble published a brief review concerning novae the same year, in a series of leaflets of the American Astronomical Society. It would be unreasonable to devote space here to this purely popularising article, were it not for the several lines of thought that we find in it. Hubble mentioned in this talk the expanding nebulae around the novae in Aquila and Aquarius. As a third similar object, Hubble mentioned the Crab nebula.

Half a century later, the outstanding Soviet astrophysicist I. S. Shklovsky wrote:

The Crab nebula played an absolutely exceptional role in the history of astronomical science. There is much truth behind the current joke that modern astrophysics can be divided into two parts: the Crab nebula physics and ... the rest. For instance, this nebula was the first identified radio source (if we ignore the Sun, of course). It was also the first identified cosmic X-ray source. It includes the shortest-period pulsar, outstanding in all respects. Finally, its continuous-spectrum optical emission has a unique nature not yet found anywhere else in space.

The fact that the nebula is located in the skies close to the bright 'guest star' of 1054 in Taurus had been noticed already in the 1920s; the 'guest' had been described in Chinese and Japanese chronicles. Astronomers had been able to determine that the nebula is expanding and measured reliably the velocity of this expansion. In one of his papers, Duncan even indicated by arrows on the photograph of the nebula the angular motions of its parts and predicted the shape of the nebula in 500 years. For some reason, however, no one before Hubble thought of projecting back in time to find out when the nebula had started to expand. Knowing the velocity of expansion, Hubble easily calculated that the nebula was born about 900 years ago, that is, at the time when the 'guest star' occurred; today we know this star as a supernova. Thus Hubble connected the Crab nebula and the supernova together not only by position in space but chronologically as well.

Hubble's conclusion may have seemed so obvious to him that he did not assign any significance to it. Never again did he mention this result; it was still too early to draw profound conclusions from the identification.

In July, Hubble and the spectroscopists Arthur King and Charles St. John represented the Mount Wilson Observatory at the third General Assembly of the International Astronomical Union in Leiden. This time Slipher, the president of Committee 28 on Nebulae, was unable to attend, and Hubble chaired the sessions. The Committee emphasised the fundamental difference between galactic (diffuse and planetary) and extragalactic nebulae. Hubble's work had removed any doubts about this point. Nevertheless, the detailed classification of extragalactic nebulae again failed to get approval, even though virtually everybody realised that there were both spiral and elliptic objects among them.

Hubble was regarded as the greatest authority on nebulae and elected as president of the Committee.

After the General Assembly, Hubble and his colleagues started on a scientific tour of the research and technical centres of the Netherlands, Germany and England. They had to think about the construction of a new, 200-inch reflector and about the necessary equipment.

In the winter of 1932, Hubble completed his latest work devoted to the Andromeda nebula. It was his farewell to the nearest neighbours of our galaxy, he was now aiming at more distant stellar systems. Actually, Hubble together with Baade and Humason discovered five to six years later 60 new Cepheid variables in the galaxies of the Local Group of galaxies, but chose not to publish these results.

Scanning the best photographs of the Andromeda nebula, Hubble noticed numerous diffuse objects. These circular objects, sufficiently concentrated but still quite distinguishable from stars, were found both against the background of the nebula and far beyond its boundaries. The brightest of them could not be confused with stars even visually through the telescope, even better so than on photographs. Humason also carefully analysed several objects at the focus of his instrument. It was possible to measure the radial velocity of one of them. It was negative, close to the velocity of the nebula itself. The general opinion was that the objects were globular clusters included in the Andromeda nebula. Hubble, however, had doubts about that and referred to the new objects as 'preliminarily identified globular clusters'. The history of science has proved that this precaution was unnecessary. At the beginning of the 1930s, however, both de Sitter and Shapley, who carefully studied Hubble's work, suspected that the new objects could be open, not globular clusters. Shapley still relied on van Maanen's results and felt uneasy because, assuming these clusters to be globular, one inevitably had to move the spiral galaxies away to distances of millions of light years.

Hubble studied carefully 140 clusters he had found: he measured their magnitudes and sizes, and characterised their degree of compactness. They resembled the clusters in our galaxy in practically all respects. Hubble only noticed that, on average, they were considerably less bright than the galactic clusters. He could not suspect then that the distance to the Andromeda nebula, which seemed to have been very reliably determined, would have to be considerably magnified in a quarter of a century.

The globular clusters thus found were seen to be concentrated at the centre of the stellar system, as they are in our galaxy. Some of them, however, were found very far away, up to two and a half degrees of arc from the centre of the Andromeda nebula. This meant that, judging by clusters, the nebula stretched to at least 30 kiloparsec. Globular clusters were also found in the northern companion of the nebula NGC 205.

Hubble discovered about a dozen such objects in the Triangulum nebula, which is located at about the same distance from us as the Andromeda nebula. He noticed their two interesting features: they were approximately one and a half stellar magnitudes weaker and the brightest of them were blue. Now we know that globular clusters had been formed in galaxies at different periods: either almost all of them formed when the galaxies themselves were formed, or they continued to form not too many millions of years ago; the former are redder, the latter are blue.

More than fifty years have passed since Hubble obtained this result. Baade doubled the number of known clusters in the Andromeda nebula even during Hubble's lifetime. Many new objects have been added also in the Triangulum nebula in recent years when research has employed Ritchey–Chretien telescopes which produce much better defined images on a wide field of view. Each subsequent author invariably mentions the contribution of the man who was the first to study the globular clusters in the galaxies of the Local Group.

The red-shift. Predecessors

Any discovery, whether minor or major, has a history. This is not accidental. 'Science is essentially sequential, systematic and possesses an "inner logic", so that each subsequent step in science stems from the preceding step' (S. I. Vavilov). This is also true for the red-shift law – Hubble's major discovery which proved that the Universe was expanding.

In 1893–4, a rich American, Persival Lowell, had a private observatory built in the Arizona desert, in Flagstaff, a town that at the time was quite small. There have been several examples in the history of science of people whose intellectual hobbies of teenage years suddenly blossomed anew in the soul of a mature person, forcing him to drop whatever he was doing and devote his entire life to science. One example is Heinrich Schliemann and his unstoppable search for the legendary Troy: he began as a salesman, banker and even a first guild merchant in Saint Petersburg in Russia, and only later became an archaeologist. Something like this happened to Lowell, who in his youth was an amateur astronomer. The lives of these two people even have quite a lot of common features: both were successful businessmen, both travelled in many countries, both made considerable fortunes and then spent the gains on their late devotion. Both decided to change their habitual way of life and did it at an age when common sense advises one that it is far too late.

The wonderful discoveries of Schiaparelli made Lowell a staunch supporter of the idea that intelligent life does exist on Mars. A new observatory was built to prove that it was true. In 1897, a sufficiently large (by the standards of that time) 24-inch refractor telescope was installed in a wooden cylindrical tower. Nowadays this instrument, with its shining brass parts, looks outmoded and even clumsy, but in its time it seemed to be perfection itself.

In 1901, Vesto Melvin Slipher, who had just received the Bachelor of Science degree from Indiana State University, joined the staff of this observatory. Lowell and Slipher had very different personalities. The former was colourful, exuberant, a master of repartee. The latter was a self-contained, thorough, withdrawn person, shunning public speeches. Although Lowell did not have an astronomer's education, he understood full well the significance of astrospectroscopy, which was still a very new field, and ordered an excellent spectrograph for his observatory.

The possibility of measuring radial velocities was regarded as a very promising field. An enthusiast in this subject who was also an amateur astronomer, Sir William Huggins, wrote, on the eve of the 20th century:

'This method of work will doubtless be very prominent in astronomy in the near future, and to it probably we shall have to look for the more important discoveries in astronomy which will be made during the coming century.'

Slipher began working with the spectrograph in the spring of 1902. Lowell was interested not only in life on Mars but also in the problem of the origin of the Solar System; several years later he suggested that Slipher try to establish the rotation of the Andromeda nebula, which at the time was regarded as a model of a planetary system in the process of formation (as postulated by the Moulton–Chamberlin model).

The conjecture was not without certain foundation. In 1888, the library of the Royal Astronomical Society in London displayed a photograph of the Andromeda nebula obtained by Isaac Roberts. The photograph revealed details that could not even be suspected by anyone who observed the nebula visually through a telescope. To some extent, it resembled Saturn: it had a bright bulge without clear-cut bounds at the centre, and more or less symmetric diffuse rings could be guessed, with a little imagination, around this core. Roberts was the first to formulate the hypothesis that the Andromeda nebula was an example of a solar system in the process of formation.

Slipher did solve the problem offered by Lowell: he measured the rotation of the Andromeda nebula. However, he made a more important discovery on the way, and his first brief note on the spectrum of the nebula holds an important place in his outstanding contribution to science. On 17 October 1912 Slipher photographed the spectrum of the Andromeda nebula with an exposure of nearly seven hours, and for the first time measured its radial velocity.

The result was so unexpected that Slipher exposed several more spectrograms before the end of the year and decided to publish the result only after reliable verification. The radial velocity was -300 km/s. The Andromeda nebula was approaching our Galaxy at a velocity which had not previously been observed for any celestial object. We do not know specifically what Slipher thought about the future when he completed writing his note, but its last words proved to be prophetic. He wrote: 'Thus extension of the work to other objects promises results of fundamental importance.'

When Slipher told Lowell about his success the latter is said to have exclaimed something like 'Looks like you've made a great discovery. Now try to test it with several more nebulae'. Professor John Miller wrote to his former student: 'It looks to me as though you have found a gold mine, and that, by working carefully, you can make a contribution that is as significant as the one that Kepler made, but in an entirely different way.' Slipher soon recorded the spectrum of the NGC 4594 nebula in Virgo. Its radial velocity turned out to be 1000 km/s.

This was the beginning of really hard work at the Lowell observatory. Even photographing the spectrum of the Andromeda nebula required an entire long autumn night. Other nebulae were much fainter, and required exposures of tens of hours. The observation of a single object took many nights and sometimes the entire moonless period. By the end of 1914, Slipher had accumulated the spectra of nearly 40 nebulae and stellar clusters; he tried to measure the radial velocities of 15 nebulae. He presented his results at the meeting of the American Astronomical Society at which Hubble was present for the first time, and published a short article about it in the January 1915 issue of *Popular Astronomy*. All nebulae moved at tremendous velocities, from two or three hundred to 1100 km/s. It was even more intriguing that almost all these velocities were positive. Slipher first thought that the signs of velocity of objects to the north and to the south of the Milky Way were different. This might signify that nebulae were flying through the Milky Way as a swarm. Further observations demonstrated that only the Andromeda nebula and its nearest neighbours in the sky had negative velocities. The average velocity of nebulae was 400 km/s, exceeding the stellar velocities by a factor of 25.

Two logical problems arose after a catalogue of the velocities of a group of objects had been compiled: to determine the velocity of the Sun relative to the group as a whole, and to try to relate the velocities to some characteristics of the objects. Slipher was too careful not to realise that he had an insufficient set of radial velocity determinations, so he refrained from the former problem. Even later, when his set of data increased considerably, he continued to treat his results on the motion of the Sun as only preliminary.

Slipher noticed, comparing the apparent elongated forms of the nebulae with their radial velocities, that ovate nebulae move faster. He had an impression that they fly through space edgeways. This was where Slipher's habitual caution left him for a while. The effect proved to be a spurious result of the small number of objects.

Less than a year after this, a short communication was sent to the same journal by Truman from Iowa State University. Its author was not well known in astronomy: he wrote only a few papers and these are of no general interest. Nevertheless, this paper of his fixed his name in the history of science as the first publication in the sequence of Hubble's predecessors in studying the motions of nebulae.

If the Sun moves through a group of objects and the components of its velocity are along three coordinate axes which are aimed at the point of vernal equinox, at the point 90 degrees from it in the equatorial plane, and at the pole (X, Y and Z), then the observed radial velocity is

$$X \cos \alpha \cos \delta + Y \sin \alpha \cos \delta + Z \sin \delta = V_r$$

where α and δ are the celestial coordinates: right ascension and declination of objects. A set of such equations can be solved for a number of objects, which gives X, Y and Z and the total velocity of the Sun, as well as the direction of its motion. This is what Truman did. There was nothing new in the method. Astronomers had applied this technique to stars for quite a long while. The new aspect was the object of study: nebulae.

In reality, the kinematic equation is not exact, reflecting only the motion of the Sun with respect to the entire group of objects while each of them also moves with respect to all others. As a result, the sought values are obtained with errors which are greater, the smaller the number of objects, the larger the spread in their velocities and the lower the accuracy of measuring their radial velocities. The uncertainty of Truman's solution was not small but the final results looked quite realistic. The Sun appeared to be moving to its apex – a point between Sagittarius and Capricorn – at a velocity of 670 km/s; this is equivalent to saying that an ensemble of nebulae was moving in the opposite direction at the same velocity.

Two Canadian astrophysicists from the Victoria Observatory, Young and Harper, experts on spectroscopic binary stars, took up the kinematic problem too, knowing nothing about Truman's work. Both the method and the original data were identical and small discrepancies were merely technical. They were ready to send their paper to the journal when Truman's work arrived. Young and Harper checked that their results practically coincided with Truman's. They called the velocity of 598 km/s, that they had determined, the velocity of the Universe.

Another paper appeared in the *Publications of the Astronomical Society of the Pacific* in the middle of 1916. Its author, the Lick Observatory assistant John Paddock, approached the problem in a slightly different way. Assume that we already know the direction of motion of the Sun (Paddock either set it equal to Young's and Harper's result or postulated other values); then the expression for the radial velocity of any nebula

can be given in the form

$$V_{\odot} \cos \lambda + K = V_r.$$

The angle λ is the angular distance on the celestial sphere between the apex of the Sun moving with respect to nebulae at a total velocity $V_{\odot}$ and the object investigated. Paddock was the first to introduce the so-called K-term for nebulae, namely, a certain increment to the solar velocity. Such a term had already been introduced for stars; its existence was detected by the Americans Frost and Adams and later confirmed by the Dutch astronomer J. Kapteyn and again by Frost in 1910. Formally, a positive K-term signified that the entire ensemble of stars moves away from us at a velocity K. It was later realised that there can be other reasons for a non-zero K-term: the effect could be caused not by a real motion but by the shift of spectral lines in the field of gravitation of massive stars or in the resultant field of gravitation of large masses of the universe (this interpretation is implied by general relativity).

Whatever version of the calculation Paddock chose, the K-term was also in the range 248–338 km/s and positive. Its sign indicated that the nebulae 'are receding not only from the observer or star system but from one another as well'. Paddock concluded that the solution obtained 'should doubtless contain a constant term to represent the expanding or systematic component whether there be actual expansion or a term in the spectroscopic line displacements not due to velocities'. The K-term for nebulae was very different to that for stellar motions, which was merely several kilometres per second.

Three papers analysing the velocities of nebulae had already been published in the USA, but the astronomer who was spending endless tiresome nights collecting the material kept his silence. Only on 13 April 1917 did Slipher present a report, 'Nebulae', to the meeting of the American Philosophical Society. The Philosophical Society brings together scientists in very dissimilar fields, so that the other two talks had no connection with astronomy. Slipher's presentation was, to a large degree, a popular review both of the general data on nebulae and of his own results. Slipher recounted the difficulties in observing nebulae; he mentioned that nebulae rotate. Slipher continued to believe that nebulae fly through space edge-forward. (Curiously, C. Wirtz continued to write about this five years later and only Lundmark closed the matter in 1925, having failed to find any correlation between the radial velocity of nebulae and their geometric shapes.)

The most important contribution of the paper lay in a different point.

Working very hard, Slipher was able by 1917 to increase the number of nebulae with measured radial velocities to 25. 'The mean of the velocities with regard to sign is positive, implying the nebulae are receding with a velocity of nearly 500 kilometres', said Slipher, adding cautiously 'that might suggest that the spiral nebulae are scattering, but their distribution on the sky is not in accord with this since they are inclined to cluster'. This argument, which in fact plays no role in this phenomenon, appeared very important to Slipher.

One could expect that now the scientist possessing this considerable amount of data would study in detail the motion of the Sun. This did not happen, however. Slipher kept speaking about such analyses as a matter for the future and made only a preliminary indication that the Sun moves at a velocity of 700 km/s towards Capricorn.

Stars surrounding the Sun did not give evidence of such motion. To Slipher, this difference was a confirmation of the idea that nebulae are separate islands in the universe. For some reason, he did not mention the work of Truman, Young and Harper.

Completing his talk, Slipher concluded unequivocally that the nebulae he had studied were definitely not objects of which solar systems similar to our system could be formed.

This was the time when World War I was raging on the fields of Europe, the Near East and the Transcaucasian region. The ties between countries, which normally were so tight, were broken and scientists of the old and new worlds knew next to nothing about what their colleagues did on the other side of the Atlantic. In fact, it was precisely at this time that most important results were obtained in Germany and the Netherlands, directly relevant to the surprising radial velocities of nebulae reported by Slipher. Albert Einstein in Berlin formulated his cosmological equation and solved it, assuming the universe to be stationary. In this solution, the hypothetical forces arising from the gravitational repulsion of the vacuum that he had introduced were balanced by the attraction of matter filling up the universe. A year later, Willem de Sitter, Professor at Leiden University in the Netherlands, which stayed neutral in World War I analysed the astronomical consequences of relativity. De Sitter found that Einstein's solution was not unique. If one assumes that the mean density of matter in the universe is very low, the Einstein repulsion forces dominate over the attraction of matter and cause its expansion. The cosmic repulsion forces are proportional to distance, and hence the velocity of mutual recession of particles of matter (individual galaxies can also be treated as particles) should be proportional to distance.

In 1916 and 1917, three papers devoted to Einstein's theory of gravitation and its astronomical applications, written by de Sitter on Eddington's proposal, were delivered to England and published in the monthly journal of the Royal Astronomical Society. One of the consequences of the war was that Slipher's list of radial velocities of nebulae did not reach de Sitter, who then knew only the results of measuring the velocities of the Andromeda nebula and of two more nebulae. He only had to point out that in contrast to the Andromeda nebula, the other objects possess positive velocities. De Sitter assumed, however, that 'spiral nebulae most probably are amongst the most distant objects we know'. He predicted with great force that '... for objects at very large distances we should expect a greater number of large or very large radial velocities'.

This was the beginning of the purely European part of analysing the motions of nebulae.

At the end of 1917, Carl Wilhelm Wirtz from the Strasbourg Observatory also introduced the K-term into the kinematic equations, knowing nothing about Paddock's work. It is very strange, but Paddock's paper was almost unnoticed. Even Hubble did not mention Paddock in a detailed description of the work of his predecessors in the book *The Realm of the Nebulae*; he assumed that the introduction of the K-term was Wirtz's achievement.

Wirtz came to the conclusion that 'relative to the present location of the Solar System taken as a centre, the system of spiral nebulae is moving away at a velocity of about 656 km/s'.

Four years later, with a doubled set of radial velocity data – 29 evaluations – Wirtz repeated his analysis and obtained practically the same result. As far as we know, it was in this paper that he gave the K-term the name 'red-shift' for the first time.

Lundmark also carried out the same calculation of the K-term in the period between the publication of Wirtz's two papers. The nature of nebulae was actually unknown at the time, so that Lundmark added planetary nebulae to the spirals and the Magellanic Clouds. However, the majority were spirals so that all versions of his solutions invariably led to the same conclusion: the K-term is very high and its sign is positive.

While Paddock, Wirtz and Lundmark were calculating the K-term, Slipher continued to measure in solitude the radial velocities of more and more nebulae. The number of nebulae with known radial velocities grew steadily and reached 45 in 1925. Nevertheless, Slipher avoided analysing the data obtained.

The world war and the breakdown of scientific communications appear to be the reasons why neither Wirtz nor Lundmark mentioned de Sitter's theory in their papers.

During the war, de Sitter not only elaborated the application of Einstein's theory to astronomy, but also accomplished another very important job which ultimately stimulated the investigation of the red-shift phenomenon.

Working in the Netherlands, de Sitter was able to receive scientific periodicals from Germany and to share scientific news with British colleagues; he thus acted as an intermediary between scientists of the two warring powers. It was he who sent Sir Arthur Eddington Einstein's paper in 1916, introduced him to general relativity and attracted his attention to one of the consequences of the theory which seemed to be verifiable experimentally. The leading British astronomer immediately realised the importance of Einstein's work and together with John Dyson (then the Astronomer Royal) started to prepare an expedition to observe the total solar eclipse of 29 May 1919. This was to be done despite the continuing war. A confirmation of one of the predictions of Einstein's theory could be obtained by photographing stars around the totally eclipsed solar disc and determining whether a ray of light passing by a gravitating body was indeed deflected.

Spurious factors may sometimes decide the fate of observing an eclipse. After months of preparations, considerable expense of time and money, and sometimes hard travel, there may be no results simply because of bad weather or an accidental cloud. Eddington's expedition was no exception. On the day of the eclipse at Principe Island off the African coast, where one of the two British expeditions anchored, a heavy shower occurred. The weather became a little better when the eclipse had already begun and the Sun was partially covered by the Moon. The astronomers had to photograph the eclipse through the clouds. Nevertheless, Eddington was able to find images of stars on several photographs. Careful measurements demonstrated that stars were indeed shifted, and shifted in agreement with Einstein's predictions. Einstein's theory was triumphantly confirmed. In December 1919 Eddington wrote to Einstein:

... all England has been talking about your theory. It has made a tremendous sensation. It is the best possible thing that could have happened for scientific relations between England and Germany.

The news that relativity theory had been tested and confirmed spread all over the world. Now other predictions of the theory were to be found; de

Sitter's papers pointed out the necessary direction of search for empirical researchers. It was necessary to test whether there is indeed a relationship between the radial velocities and the distances to remote objects.

Wirtz was the first to respond to this challenge. In spring 1924, he published a paper 'De Sitter Kosmologie und die Radialbewegungen der Spiralnebel'. But how could the distances to nebulae be obtained? Indeed, at that time Hubble had not yet measured the distances even to the nearest nebulae – in Andromeda and Triangulum. Wirtz decided to choose the apparent diameters of nebulae as a measure of distances, assuming that their true dimensions are on average identical. In this assumption, the apparent diameter of a nebula is smaller, the further away the nebula is. The sought relation between the apparent diameter and velocity, or rather a hint at such a relation that Wirtz found, was that the radial velocity was greater the smaller the nebula he chose. But the formula obtained by Wirtz was not quite the one predicted by the theory. De Sitter showed that velocity and distance must be in a linear relationship; Wirtz decided to take not the diameters of nebulae but their logarithms, possibly in order to compensate somehow for the shaky foundation of his hypothesis. Another factor reduced the reliability of his result. It was found that the surface brightness of nebulae also correlated with velocity. The velocity was higher in compact nebulae. It was not understood at the time that this was an effect of selection, which is so often encountered in astronomy. It meant merely, that, among generally faint objects, those nebulae were selected for observations whose surface brightness was higher, especially at the centre. Then their spectra could be photographed over acceptably long exposure times.

A similar study was completed in summer of the same year by Lundmark. He faced the same difficult problem of distances to nebulae. Naturally, he again had to resort to the hypothesis of identical sizes. Lundmark added to it an additional assumption that all nebulae have identical luminosities. Combining the two approaches, Lundmark calculated the distances to all nebulae in relative units. For the unit of distance, he chose the distance to the Andromeda nebula. However, he also failed to arrive at a reliable, conclusive result. Lundmark concluded: 'Plotting the radial velocities against these relative distances, we find that there may be a relation between the two quantities, although not a very definite one'.

Nothing decisive was produced in the next year, 1925. At last an American astronomer at the Mount Wilson Observatory, Gustav Strömberg, joined the work on studying the motions of nebulae. However, he was using the same set of data on radial velocities and the same hypothesis on the apparent brightness of nebulae as a measure of distance. As before, he found no more than a hint at a relation between velocity and distance. Strömberg concluded in a precise, if disappointed, manner that 'We have found no sufficient reason to believe that there exists any dependence of radial motion upon distance from the Sun'.

By the time Strömberg's work had been completed, tireless Lundmark had published a new study. This time he made an attempt to represent the red-shift effect in kinematic equations, not by an ordinary K-term but by an expression with a constant term and two terms containing distance to first and second powers, respectively. The values of the sought coefficients were very uncertain. However, since the coefficient with squared distance was found to be negative, Lundmark concluded that 'one would scarcely expect to find any radial velocity larger than 3000 km/s among the spirals'. In fact, this threshold was left behind before five years had passed.

The last, and essentially unsuccessful attempt to establish a relation between velocity and distance to nebulae using their apparent diameters was made by the German astronomer A. Dose in 1927.

Hubble's Law

Any serious investigator was already convinced that the problem was not in the small number of measured radial velocities or in their insufficient accuracy, but in the degree of uncertainty in the evaluations of distances to nebulae. The key to solving this cardinal problem was in Edwin Hubble's hands. He had read the papers of his predecessors and undoubtedly expected that there must be a relation between velocities and distances to nebulae.

Hubble knew at least one theoretical paper which predicted how the red-shift should be related to the distance to galaxies. He had already discussed de Sitter's relativistic model of the universe in 1926, in his paper 'Extragalactic Nebulae', and very probably thought about testing the predictions of theorists, even though he was never very enthusiastic about theories.

Cosmological models based on general relativity were completely developed by the end of the 1920s. However, they remained either quite unknown to astronomers or produced no response from them. In all likelihood, there were several reasons why a theoretical prediction of the most important natural phenomenon left those who could test this prediction completely indifferent. At the first stage only Russell and Shapley, as far as we can judge, discussed in their correspondence a relationship between de Sitter's theory and the possible velocity–distance relation for spiral nebulae and even globular clusters which at that time seemed to be almost as distant.

The first reason was, presumably, that the cosmological models were based on Einstein's general relativity theory, which is mathematically quite complex and, furthermore, involves revolutionary new concepts of space and time and new principles of gravitational interaction. In that epoch, not only observing astronomers but even theoretical physicists had the greatest of difficulties in mastering the new ideas and making them their own, that is, understanding them thoroughly and learning to apply them to specific studies. The first reason thus stemmed from the complexity of the theory and poor communications between theorists and observers. The second reason was psychological, possibly rooted in the very unusual conclusions of the theory which stated, for example, that space may be closed or that our world could have started to evolve at a certain time in the past. Practical astronomers who penetrated deeper and deeper into cosmic space by their new powerful telescopes tended to reject psychologically the reality of statements which profoundly changed their picture of the universe. Note that this situation repeated itself forty years later in the story of the prediction and discovery of the microwave background radiation of the hot universe.

But let us return to the early 1920s. In 1922 and 1924, the Russian mathematician A. A. Friedmann derived and completely solved the cosmological equations. These equations were implied by Einstein's theory and described the general structure and evolution of the universe; Friedmann assumed that matter is distributed uniformly on very large scales and that all directions in space are equivalent. The main conclusion from Friedmann's solutions was that in the general case, matter in the universe cannot be stationary on average on a large scale: the universe must either expand or contract. This conclusion was derived in a strict mathematical way; however, it is essentially quite simple.* The only forces which act in a uniform universe are the gravitational forces. Therefore, if we imagine that the enormous masses of the universe are at some moment

---

* It is important to know that the simple interpretation of Friedmann's main conclusions as given below was not immediately understood.

on average at rest relative to one another, the next moment gravitation
will set them in motion and matter will start to contract. Galaxies can
be regarded as the 'particles' of this matter. Of course, the universe
need not necessarily contract. If all masses are given certain velocities at
which to recede from one another, the universe will expand, with gravitation only slowing down this expansion. Only the initial conditions decide
whether expansion or contraction will dominate; in other words, the processes which fix the initial velocities of the masses are decisive. Actually,
Einstein had introduced into these equations one more type of force: the
putative forces of gravitational repulsion of the vacuum. These forces
were to be weak and to manifest themselves only at very large cosmological distances. Einstein introduced these forces specially to construct a
static model of the universe which neither expands nor contracts. In this
solution, the attractive gravitational forces are balanced out by repulsive
forces. Friedmann's equations also take into account the Λ-term. The
forces of repulsion that this describes attenuate the gravitational forces
of matter.* However, a special choice of initial conditions is needed to
balance the forces exactly and so obtain Einstein's model. This model,
suggested in 1917, is a particular case of Friedmann's model. Another
particular case is de Sitter's model in which any gravitating matter is
absent and vacuum repulsion dominates.

It should also be added that Friedmann's equations describe not only
the dynamics of the motion of masses in the universe but also the
geometric properties of space or, in the parlance of relativity theory, the
degree of curvature of space, which varies as the universe expands.

Friedmann's first paper proving that the universe is not static was
received by the very well-known German journal *Zeitschrift für Physik*
at the end of 1922. Einstein was so sure that his model was correct,
he believed so much that the solution of the cosmological equations
he needed was static, that he decided that Friedmann's paper was in
error. By the middle of September 1922, Einstein's brief response was
received by the *Zeitschrift für Physik*. Professor V. A. Fock remarked
that in this paper Einstein 'informs in a slightly haughty manner that
Friedmann's results looked suspicious and that he had found an error
whose correction reduces Friedmann's solution to the stationary one'.

Einstein's opinion was communicated to Friedmann in a letter from
Yu. A. Krutkov, his colleague in Petrograd, who at that time was in the

---

* One may choose a negative Λ-term; then it describes additional attractive forces in the
vacuum. We will not analyse these possible alternatives here.

Netherlands. In December 1922, Friedmann wrote a letter to Einstein, in which he explained in detail the essence of his calculations and tried to prove that he was right. The letter ends with the following words:

If you find the calculations outlined in my letter to be correct, I would be very much obliged if you informed the editors of *Zeitschrift für Physik* about it; it would be fine if in this case you sent to the journal an addendum to your earlier opinion and suggested that a part of this letter of mine be reprinted together with the addendum.

Einstein did receive the letter (it was found in his archive) but he either failed to read it in time or did not pay attention to it, being quite sure of his result.

In March 1923, Krutkov met Einstein in Leiden, at the house of the well-known Dutch physicist Paul Ehrenfest and succeeded in proving to Einstein in a number of discussions that the Russian mathematician was right. Krutkov wrote in his letter to his sister on 18 May 1923: 'Victory over Einstein in the debate about Friedmann. Petrograd's honour has been saved!'

Immediately after discussions with Krutkov, Einstein sent the following short note to the *Zeitschrift für Physik*:

On A. Friedmann's paper 'On the Curvature of Space'. In my previous note, I criticised the above-mentioned work. However, as I found from Friedmann's letter communicated to me by Mr Krutkov, my criticism was based on an error in calculations. I believe that Friedmann's results are correct and shed new light. It is found that the field equations allow, in addition to a static, also dynamic (that is, variable with relation to time) centrally symmetric solutions for the structure of space.

These words written in 1923 mark the fundamental importance of Friedmann's theoretical discovery of the non-stationary universe. However, later events demonstrated that even though Friedmann's paper had been published in a widely read journal and although Einstein himself recognised it, the paper was not noticed, not only by astronomers but also by theoretical physicists. It is not easy to explain why this happened.

In 1923, the German mathematician Hermann Weyl pointed out that if galaxies were placed at a low density into de Sitter's empty universe where only vacuum repulsion forces act, so that the gravitational attraction of galaxies is negligible in comparison with the repulsive forces described by the Λ-term, the galaxies start to move at velocities proportional to distances between them (while distances are still relatively small).

Another theoretician, H. Robertson, came to the same conclusion in 1928. Furthermore, Robertson compared the distances calculated from

Hubble's data of 1926 with the velocities obtained by Slipher and found an approximate confirmation of the law of proportionality of velocity to distance. We do not know whether Hubble knew about Robertson's results at the time; it would be interesting for the history of science to clarify this point.

In 1927 J. Lemaitre, a student of Arthur Eddington, essentially repeated Friedmann's work. As did Friedmann before him, Lemaitre also concluded that the universe was non-stationary. Lemaitre also derived a linear relation between velocity and distance (for small distances); this relation actually reflects the uniformity of the universe. The proportionality coefficient found in his work was not very different from that soon obtained by Hubble.

Actually, neither Hubble himself at the beginning of his work nor other direct participants in the first discussions of his discovery knew or remembered all the theoretical results. In all likelihood, de Sitter's model with the predicted recession of galaxies in an almost empty universe and Einstein's stationary universe model were the only schemes that were considered at that time.

The fateful hour was approaching in Hubble's life. As a first step, he wanted to make sure that the radial velocities of increasingly distant nebulae continue to grow. Slipher's modest telescope was not equal to this task. The problem could be solved only at the Mount Wilson Observatory with its largest-in-the-world instrument operated by the expert observer Milton Humason.

About a year before the main accomplishment of his life, Hubble composed a list of very faint and, presumably, very distant nebulae whose radial velocities should be measured; at this time, Humason photographed the spectrum of the NGC 7619 nebula – a member of the Pegasus cluster. Night after night, Humason locked the slit of the spectrograph on one and the same object so as to accumulate the necessary exposures of 36 and 45 hours. Finally, the spectrum was obtained and measured. Absorption lines in the spectrum of the nebula were found to be shifted towards the red end of the spectrum and its radial velocity was evaluated as 3779 km/s. This was a success that Hubble impatiently expected.

Now he had to move fast. On 17 January 1929 two small papers were received by the *Proceedings of the National Academy of Science*. One was Humason's, reporting the measurement of the radial velocity of NGC 7619. Another was Hubble's, also short – only six pages – 'A Relation Between Distance and Radial Velocity Among Extra-Galactic Nebulae'.

This brief article must be regarded, together with very few others, as one of the most prominent papers in the entire history of astronomy. The issue of the proceedings appeared in March. Here are the words with which Hubble began his paper:

Determinations of the motion of the Sun with respect to the extragalactic nebulae have involved a *K*-term of several hundred kilometres which appears to be variable. Explanations of this paradox have been sought in a correlation between apparent radial velocities and distances, but so far the results have not been convincing. The present paper is a re-examination of the question, based on only those nebular distances which are believed to be fairly reliable.

Hubble possessed a set of 46 measurements of radial velocities, from negative velocities of the nearest galaxies (the Andromeda nebula, its companions and the Triangulum nebula) to positive, reaching 1000 km/s for the most distant objects. We are inclined to think now that Hubble had a gift of prophesy and knew precisely what he was going to need in several years. It was now time to make use of some important results of 1926: the luminosities of the brightest stars in nebulae were found to be approximately equal. The luminosities of the galaxies that Hubble studied himself were also not very different. Radial velocities of 24 nebulae at known distances from the Sun had already been measured. Four of them, those with the largest velocities, were found in the Virgo cluster.

No matter whether the motion of the Sun was taken into account or ignored, a simple comparison of velocities and distances demonstrated with certainty that a relation does exist (Figure 2). Hence, a *K*-term must indeed be introduced into the kinematic equation but it must be proportional to distance, *Kr*. The solution of the equations implied that the Sun moved at a velocity of about 280 km/s in the direction of Lyra, towards Vega, a bright star in the northern skies. The principal result was that the numerical coefficient *K* in the product *Kr* was calculated quite reliably. In two slightly different solutions, it was found to be approximately +500 km/s per megaparsec. This result signifies that nebulae fly away from us and from one another too, and that their velocities indeed increase linearly with distance. As for the remaining 22 nebulae whose motion along the line of sight was known but the data on distance remained unknown, they could be approached in different ways. First, one could try to evaluate the distances to them using their apparent magnitudes. This is what Hubble did, and his kinematic result remained practically unaltered: +530 km/s per megaparsec. The problem could be inverted: one could regard the dependence of velocity on distance to be

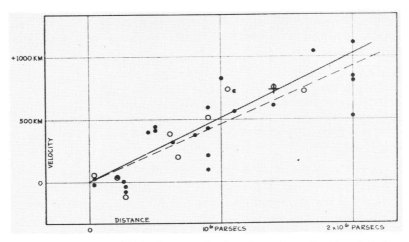

Figure 2. Hubble's diagram: the first dependence that he plotted
of the radial velocities of galaxies and distances to them (1929)

known, calculate the distances given by it, and then evaluate the absolute
magnitudes of the nebulae. On average, the absolute magnitude was equal
to -15.3^m, that is, to the value that Hubble had obtained for nebulae at
already known distances. There were no contradictions. Hubble's main
conclusion was: 'The results establish a roughly linear relation between
velocities and distances among nebulae for which velocities have been
previously published. In order to investigate the matter on a much
larger scale, Mr Humason at Mount Wilson has initiated a program of
determining velocities of the most distant nebulae that can be observed
with confidence'. Indeed, the radial velocity of the first distant galaxy on
Hubble's straight line corresponded to a distance at which its absolute
stellar magnitude coincided with the magnitudes of the brightest members
of clusters of galaxies. At the end of his paper, Hubble lucidly formulated
the profound meaning of his result: 'The outstanding feature, however,
is the possibility that the velocity–distance relation may represent the de
Sitter cosmology, and hence, that numerical data may be introduced into
discussions of the general curvature of space'.

Hubble denoted the proportionality coefficient in this relation by a
letter K. Later the coefficient was named in honour of Hubble, as the
Hubble constant; it is denoted by H. Presumably, the first to refer
to the proportionality of the red-shift to the distance to galaxies as
'Hubble's Law' was the Caltech Professor Richard Tolman, a very well-

known theoretician and specialist on applications of general relativity to astronomy. The title of his paper was 'On the astronomical implication of the de Sitter line element for the Universe'. In this paper, Tolman referred to the red-shift versus distance dependence as a well-established fact in the world of galaxies. Incidentally, Tolman's paper is dated 25 February 1929 while Hubble's paper appeared only in the journal issue of 15 March. It seems that Tolman already knew about the result before the publication. This is not really surprising: the office of the Mount Wilson Observatory is in Pasadena, the same town where Caltech is located, and the two scientists knew each other well.

After several years had passed, Milne and other astronomers were writing about the relation as the generally familiar Hubble's Law.

In Russia, it was again V. Maltsev who was the first to describe the new outstanding achievement of Hubble in the pages of a popularising journal for everyone interested in science. However, his presentation is rather fuzzy since he evidently had not read Hubble's original paper.

At the end of June, at the meeting of the American Astronomical Society at Berkeley, Hubble reported the results of his attempts at measuring the distance to a large cluster of galaxies in Coma Berenices. Here a thousand nebulae were seen in an area of about ten lunar discs; the brightest of them were observed visually about 70 years before Hubble. If one assumed for these galaxies the average absolute magnitude evaluated by Hubble, the distance to the cluster was found to be 16 million parsec, or seven times the distance to the Virgo cluster. The light from these galaxies had taken 50 million years to reach Earth.

In Spring 1929, Humason and Pease overcame the greatest difficulties and measured the radial velocities of three nebulae in Coma Berenices. These nebulae were considerably fainter than those that Humason observed in Virgo. Their velocities lay around the value 7500 km/s, with an error of 1200 km/s.

Russell, who always responded immediately to all important news in astronomy, wrote then in *Scientific American*:

The existence of the strange 'distance effect' appears therefore to be conclusively proved. If it continued to be proportional to the distance for still more remote nebulae, their spectra would be very strange. A nebula ten times as far away as Humason's [NGC 7619] could still be easily photographed as an object of the 17th magnitude, and for this we might expect a velocity of nearly 40 000 km/s. This would shift the sodium lines from the yellow part of the spectrum far into the red!... The reader has probably long been asking, 'What does this all mean? Are the nebulae really flying out in all directions – away from us and therefore from one another – so that the universe of nebulae is expanding...' The best answer that has yet been suggested comes from a peculiar form of the theory of relativity suggested a few years ago by the great Dutch astronomer de Sitter... If originally they [nebulae] were fairly close together, they would after the lapse of ages, be receding from one another with speeds proportional to their distances, just as Hubble's investigations indicate. It would be premature, however, to adopt de Sitter's theory without reservation. The notion that all the nebulae were originally close together is philosophically rather unsatisfactory. One asks why, and finds no answer.

The summer of 1929 came and Hubble departed, as usual, to spend his holidays in the Sierra Nevada Mountains; when he came back, he found a package with a new May issue of the *Bulletin of the Astronomical Institutes of the Netherlands* with de Sitter's paper 'On Stellar Magnitudes, Diameters and Distances to Extragalactic Nebulae and Their Apparent Radial Velocities'. This was a detailed and fundamental paper about precisely the subject that had occupied Hubble for several months already. However, it did not contain, and could not contain, anything new from an observer's standpoint. De Sitter had no new data on radial velocities and he was taking stellar magnitudes from the published literature. As for the estimates of distances, de Sitter combined them from Hubble's recently published data, from Lundmark's paper of 1927 and from a small number of other sources. Naturally, de Sitter obtained practically the same results as Hubble did. The only thing which can be said to make this paper different, was de Sitter's wish to find immediately the theoretical implications of the expansion of the observable universe. This aspect attracted special interest from Hubble, an observing astronomer; however, the hastiness of a theorist's approach clearly irked his ego and he had to state firmly in his letter to de Sitter:

There are some points in Bulletin 185 which I venture to call to your attention. The possibility of a velocity–distance relation among nebulae has been in the air for years – you, I believe, were the first to mention it. But our preliminary note in 1929 was the first presentation of the data where the scatter due to uncertainties in distances was small enough as compared to the range in distances, to establish the relation. In that note, moreover, we announced a program of observations for the purpose of testing the relation at greater distances – over the full range of the 100-inch, in fact. The work has been arduous but we feel repaid since the results have steadily confirmed the earlier relation. For these reasons I consider the velocity–distance relation, its formulation, testing and confirmation, as a Mount Wilson contribution and I am deeply concerned in its recognition as such... We have always assumed that, where a preliminary result is published and a program is announced for testing the result in new regions, the first discussion of the new data is reserved as a matter of courtesy to those who do the actual work. Are we to infer that you do not subscribe to this ethics; that we must hoard our observations in secret? Surely there is a misunderstanding somewhere.

De Sitter was somewhat annoyed by the tone Hubble took in his letter but their future correspondence remained friendly. De Sitter completely accepted Hubble's attitude.

The theoretical interpretation of Hubble's discovery in the framework of de Sitter's model met with obvious difficulties since the model postulated a negligibly small mean density of matter in the universe. In fact, the distribution of galaxies in space showed that the density is by no means low. At that period, Lemaitre wrote a letter to Sir Arthur Eddington, recalling to his attention his own paper of 1927 about the expansion of the universe at a finite (but not negligibly small) mean density of matter. Eddington immediately realised that Lemaitre's model gave a theoretical explanation of Hubble's conclusions. He passed the news to de Sitter and the two famous theoreticians warmly greeted Lemaitre's work. Friedmann was already dead by that time and, unfortunately, nobody recalled his pioneering and exhaustive papers. This is how Lemaitre was announced to be 'the father of the theory of the expanding universe'. He drew a lucky number. Only many years later was Friedmann's precedence gradually recognised; today it is firmly established.

In the meantime, the observations of Hubble and other astronomers were bringing new results. Walter Baade discovered a fairly large cluster of galaxies in Ursa Major while he was working in Hamburg, before moving to the USA. The members of this cluster were much fainter than in the Virgo cluster, and Hubble daringly predicted that their radial velocity must be around 12 000 km/s. Humason's observations of one of them gave the figure 11 800 km/s!

Then W. Christie, photographing Neptune in search of its satellites, found another cluster of faint nebulae in Leo, not far from Regulus. The brightest of them, a small, slightly diffuse 'star' on a photograph made on the 100-inch, was fainter by an entire magnitude than the brightest nebula in the Ursa Major cluster. Hubble and Humason realised that the Leo cluster might prove to be the farthest object known so far in the universe. They specially modified their equipment so as to be able to photograph the spectrum of the nebula at feasibly long exposures. Finally, they obtained the result: 19 700 km/s. A lens of modest magnification was sufficient to see on the spectrogram, without special measurements, that the familiar H and K lines of calcium were shifted appreciably towards the red end of the spectrum.

Two years of intense work had passed. The time had come to summarise the results. By March 1931, Humason and Hubble had prepared two most important papers. They were soon published together in the

Astrophysical Journal. Now Humason could present the radial velocities of 46 nebulae, both of isolated objects and also of members of groups and clusters in Virgo, Pegasus, Pices, Cancer, Perseus, Coma Berenices, Ursa Major and Leo, the distances to which had already been evaluated by Hubble.

The new joint paper by Hubble and Humason on the determination of the dependence of nebula velocity on distance was a solid and serious investigation. Using not very distant nebulae containing known Cepheids as a basis, Hubble and Humason confirmed again that the brightest nonvariable stars in nebulae are of roughly the same luminosity and can be used as distance indicators. However, this method ceased to work when nebulae are so far away that individual stars in them become unresolvable. However, the absolute magnitudes of the nebulae themselves can be considered as equal, under a certain approximation. Once this assumption is made, a measure of the distance to a nebula is its apparent brightness.

The paper of Hubble and Humason contained quite a few details which are not very interesting now. However, if we ignore them, the main message of the paper was a new list of nebulae with measured radial velocities and apparent magnitudes. Now it was possible to start testing whether the relation between velocity and distance survives for considerably more remote clusters of nebulae: 32 million parsec from us, by Hubble's estimate.

The conclusion was unambiguous:

The observations now cover a range about eighteen times that available for the preliminary investigations and approach the limit of present instrumental equipment; but the form of the correlation is essentially unchanged,... and hence the velocity–distance relation appears to be a general characteristic of the observable region of space. Aside from its cosmological significance, the relation offers a new method of determining distances of individual objects in which the percentage errors actually diminish with distance. This opens new possibilities for the investigation of nebulae....

Even the Hubble constant remained almost the same, having been changed from 500 km/(s.Mpc) to 560 km/(s.Mpc).

One has to admit, though, that Hubble was unable to become an authority on the general implications of his discovery. He was not a theoretician, only an observer. It is also possible that he was deeply impressed by the hypothesis proposed by F. Zwicky of Caltech, that the red-shift is not a consequence of motion but the result of energy loss by light quanta on their long passage through space. This is a likely reason for Hubble and Humason to start emphasising that they deal only with the observable, 'apparent' velocities of galaxies. In 1931 Hubble wrote to de Sitter:

Mr. Humason and I are both deeply sensible of your gracious appreciation of the papers on velocities and distances of nebulae. We use the term 'apparent' velocities in order to emphasize the empirical features of the correlation. The interpretation, we feel, should be left to you and the very few others who are competent to discuss the matter with authority.

Three years later, Humason had measured the radial velocities of 35 new nebulae not included in clusters. Now the researchers knew the radial velocities of 85 nebulae and could again test the velocity–distance relation, in other words, a relation between the apparent magnitude as a measure of distance and the logarithm of the radial velocity.

Isolated nebulae yielded essentially the same dependence as nebulae in clusters. The only difference was that the curves were shifted relative to each other; this effect was easy to explain.

The relation between the velocities of nebulae and distances to them was again confirmed. Could it be extended to greater distances? Many astronomers asked this question. New observations were needed, and Humason wrote:

An attempt will be made to extend the observed range in distance by measures of fainter clusters of nebulae. Some extension seems quite possible, but the limit with the 100-inch reflector will be reached at about photographic magnitude 17.5. Exposures necessary for the fainter nebulae are not as long as the magnitudes would indicate because the red-shift is so large that the H and K lines are brought into the region to which the photographic plate is very sensitive. Further, lower dispersion can be used, or, since the red-shift is larger, a larger probable error can be tolerated. The main difficulty arises from the fact that at the photographic magnitude 17.5± the nebulae become so faint visually at the Cassegrain focus of the 100-inch reflector that they cannot be seen on the slit of the spectrograph.

While Humason was photographing the spectra, Hubble was discovering new, more distant clusters of galaxies one after another: in Gemini and Corona Borealis whose brightest objects were of almost 17th stellar magnitude, and in Bootes where the brightest nebula is another magnitude fainter. In 1931, Walter Baade found another cluster, almost as faint, and again in Ursa Major. The sizes of these clusters were now not several lunar diameters but a small fraction of one diameter.

In such distant nebulae, which leave on a photographic plate only a short faint spot, it is absolutely impossible to detect any distance indicators, not only Cepheids but any brighter supergiant stars, novae, or globular or open clusters. It was necessary to use the apparent stellar magnitudes of the nebulae themselves, assuming that the spread of their true luminosities was sufficiently small. Hubble wrote: 'The general distance criterion which may be applied throughout the observable region of space is furnished by total luminosities, or, more precisely, the luminosity function of nebulae'. He tackled the nebular luminosity problem again and solved it in two ways. Now the radial velocities of more and more remote nebulae were needed to test Hubble's Law.

In 1936, Humason published data for 100 nebulae, including members of clusters in Corona Borealis, which was found to recede at a velocity of 21 000 km/s. For Gemini, measurements gave the velocities of 23 000 and 24 000 km/s, and for Bootes, 39 000 km/s. Humason measured the record velocity of 42 000 km/s for one of the nebulae in the Ursa Major cluster. Hubble's Law continued to hold.

But this was already the limit. When Hubble found on his plates a cluster in Hydra, more remote and fainter than all previous ones, Humason was unable to measure the radial velocity of even the brightest nebula of the cluster.

The 100-inch reflector had thus completely exhausted its potential in measuring radial velocities. At the same time, the problem of the origin of the red-shift in galactic spectra remained unanswered. It was necessary to understand whether the red-shift is indeed caused by the Doppler effect due to the expansion of the universe, or by some other yet unknown physical effect, for instance, by 'ageing' of photons during their protracted journey through space. If the expansion predicted by the cosmological theory was a reality, it was necessary to measure the relativistic effects and calculate the parameters of the cosmological model.

One model sets itself apart by its properties among other models of the expanding universe. First of all, it assumes that the Λ-term is zero, that is, that there is no repulsion (which was specially introduced by Einstein to construct the theory of a static universe). Models without the Λ-term divide into open and closed ones. In models of the former type, the density of matter in the universe is low and the gravitational forces are unable to stop completely the expansion of matter: the expansion lasts indefinitely. In the latter models, the density is high, the gravitational forces are strong enough to stop the expansion and cause the universe ultimately to contract. Closed models imply closed space, while the space of open models is infinite and obeys Lobachevskian geometry. The threshold value of the mean velocity of matter in the universe, separating these two cases, corresponds to the critical density. It is determined by the Hubble constant, and equals about 5×10^{-28} g/cm^3 for $H = 500$ km/(s.Mpc). A model in which the density equals the critical density is special in that its three-dimensional space is described by Euclidean geometry.

In the expanding-universe models, distances between galaxies in the past were smaller than today, and the mean density was higher than we find it now. Hence, the velocity of mutual recession of galaxies was higher and we inevitably come to a conclusion that at some moment in the past the density was infinitely high. (Neither galaxies nor individual heavenly bodies could exist at that time, they should have evolved later in the course of expansion of the universe.) This moment of the formally infinite density of matter, the moment when the expansion began, is known as the cosmological singularity. At the cosmological singularity, the 'Big Bang' began which imparted the initial recession velocities to the matter in the universe.

How far back could that happen? It is not difficult to evaluate this time. If two galaxies constantly receded from each other at a constant velocity, then the time when they were at a common point can be obtained by dividing the distance between them by their velocity. In view of Hubble's Law, $V = Hr$, we find that this time interval equals $1/H$, independently of distance. Therefore, if the recession velocity of each galaxy were not gradually diminished by gravitation, they would all be at the same spot at the moment of time $1/H$. In fact, velocities were greater in the past. However, if the density of matter in the universe is not very much higher than the critical value, and this is undoubtedly true, this deceleration would not change our estimate within an order of magnitude. With the value of H found by Hubble, we obtain the time of evolution: $1/H \approx 2 \times 10^9$ years.

The age of our planet was estimated at the end of the 1920s and the beginning of the 1930s on the basis of the radioactive decay of uranium in the crust as from two to six billion years. Rutherford also found that, judging by the relative abundances of uranium isotopes 235 and 238 in rocks, the age of the Earth was about three billion years. In 1930 Eddington noticed that the time $1/H$ is very close to the age of radioactive elements and differs greatly from the estimated age of stars. At that stage, stars were believed to have lived for much longer, about a thousand billion years. This followed from the assumption that the source of stellar energy is the conversion of their mass into radiation. It was assumed that practically all mass can transform into radiation obeying Einstein's formula $E = mc^2$. Additional arguments in favour of this long stellar lifetime followed from Jeans' estimates of the duration of dynamic processes in stellar systems.

That was the famous contradiction of two time scales. Indeed, if stars live for hundreds of billions of years, their age must be much greater than the age of the universe!

How could these very different estimates be reconciled?

Cosmologists tried to 'stretch' the time of expansion of the universe, assuming that the Λ-term is, after all, non-zero. On the other hand, by the end of the 1930s it became clear that the source of stellar radiation is nuclear energy. Only a small part of the entire mass of a star transforms into radiation, so that the estimated lifetime of stars had to be reduced by two orders of magnitude. At the same time, more thorough analysis of galaxies made it necessary to reject Jeans' arguments about extremely long lifetimes of stellar systems. Some time later, the estimate of the duration of expansion of the universe changed too, because it was found that the value of H determined by Hubble was greatly overestimated. Finally, all apparent contradictions between these different 'cosmic scales' have evaporated, even though some questions still survive. In World War II, Hubble looked over his and his colleagues' achievements of the last decade and formulated the relationship between theory and observations in this way:

Mathematics deals with possible worlds, with the infinite number of logically consistent systems. Observers explore the one particular world we inhabit. Between stands the theorist. He studies possible worlds but only those which are consistent with the information furnished by the observer. In other words, theory attempts to segregate the minimum number of possible worlds which must contain the world we inhabit. Then the observer, with new factual information, reduces the list still further. And so it goes, observations and theory advancing together toward the common goal of science, the structure and behavior of the physical universe in which we live.

Hubble as observer was searching for tests that could enable scientists to understand the fundamental properties of the universe. One of them looked the most promising to him.

However, here we have to turn back some pages of Hubble's biography and return to 1926. By that time Hubble had thought of conducting a statistical survey of nebulae for the entire sky. Numerous photographic plates of the sky with thousands of nebulae had by that time been accumulated at the Mount Wilson Observatory. However, they were exposed under different conditions and did not cover the sky uniformly, and were thus unsuitable for solving the problem thus formulated. Hubble started systematic photographing under standard conditions using the 60- and 100-inch reflectors. His plates contained about 60 000 nebulae. Hubble counted 44 000 objects on 1283 plates, carefully scanning them at least three times under high and low magnification. This was a tremendous task, requiring untiring concentration and hard work. The next step was equally labour consuming: introduction of necessary corrections into counting results.

We observe distant nebulae through our stellar system. Nebular counts tell us very much about the world of galaxies beyond our own galaxy but also inform us about some characteristics of the latter. Hubble was solving two such problems.

He did not find any nebulae along the Milky Way band. A clear-cut zone of avoidance was apparent there, the dust in our galaxy completely blocking out very distant objects; further out, there were zones with partial absorption of light, and then the area of normal distribution of nebulae.

Hubble's counts demonstrated that the entire sky must contain up to 75 million nebulae accessible to the 100-inch telescope. Hubble's main conclusion was that the number of nebulae with magnitudes (corrected for the red-shift effect) from bright to increasingly faint increases in a way predicted for the uniform distribution of nebulae in space. This gave the most important characteristic that the mean matter density in space was equal to about 10^{-30} g/cm^3.

Hubble first reported the results of his work in summer 1931, at the meeting of the American Astronomical Society at Pasadena, but published the final paper only three years later. This was an outstanding paper both in the sheer amount of material collected over numerous sleepless nights and in the thoroughness of processing the material. Hubble had predecessors in nebula counts but no one had dared to go to still fainter objects for more than half a century.

In 1935, Hubble and Tolman suggested two ways to study the nature of the red-shift, avoiding the measurement of radial velocities. 'The possibility that the red-shift may be due to some other cause, connected with the long time or distance involved in the passage of light from nebula to observer, should not be prematurely neglected'. In conclusion they stated that they 'both incline to the opinion, however, that if the red-shift is not due to recessional motion, its explanation will probably involve some quite new physical principles'.

The first method is essentially simple: it was shown that the distribution of brightness in an elliptical nebula depends on the origin of the red-shift, that is, whether it implies the actual receding of galaxies or a manifestation of as yet unknown factors. This is where Hubble could use his old paper on brightness of galaxies. However, difficulties that proved insurmountable at the time blocked any progress in the application of the method.

Hope focused on the second method. The theory gives mathematical expressions for the relation of the number of galaxies to their stellar magnitudes. However, the red-shift attenuates the light of galaxies and certain corrections have to be introduced to their measured values. Two effects are responsible for this. First, owing to the red-shift, each photon reaching the observer possesses lower energy; Hubble called this phenomenon the 'energy effect'. Secondly, if a galaxy actually recedes, the photons it emits arrive at the observer less frequently than if the galaxy were stationary: the 'number effect'. If the red-shift were caused by 'ageing' of photons on the way to the observer, there would be no number effect.

The application of this method involves enormous difficulties. For instance, light of distant galaxies comes to us reddened, so that another region of their spectrum falls into the range of sensitivity of photographic plates. This means that large corrections need to be made, and they are not well known. But the principal difficulty lies in the fact that when we observe distant galaxies we see the light they emitted in the past when the properties of radiation sources (their luminosity, size and so forth) could be different.

Hubble clearly understood the seriousness of the difficulties he was facing and nevertheless decided to compare the theory with galaxy counts. He conducted new counts of galaxies both brighter than those counted before and fainter than before, lying at greater distances; he was able to move down another unit of magnitude. The hope was that now that all the necessary corrections had been taken into account, all the effects due to the space surrounding us must manifest themselves even better. In fact, the results were very discouraging. Hubble had to acknowledge:

The observations may be fitted into either of two different types of universe... If red-shifts are not velocity shifts, the apparent distribution agrees with that in an Einstein static model of the universe or an expanding homogeneous model with an inappreciable rate of expansion... If red-shifts are velocity shifts which measure the rate of expansion, the expanding models are definitely inconsistent with the observations unless a large positive curvature (small, closed universe) is postulated. The maximum value of the present radius of curvature would be of the order of 4.7×10^8 light years; and the mean density, of the general order of 10^{-26}. The high density suggests that the expanding models are a forced interpretation of the observational results.

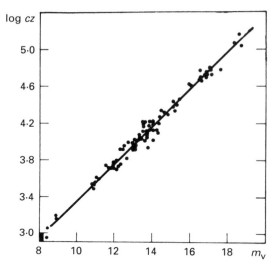

Figure 3. The apparent stellar magnitude of the brightest galaxies in 82 clusters as a function of the red-shift z. The black rectangle in the lower left-hand corner represents the range of data available to Hubble in 1929. The figure is taken from the paper of Sandage and Tammann of 1981, based on earlier measurements by Sandage.

This collision with an unsolvable contradiction clearly showed to Hubble that all hopes of progress rested with the new 200-inch telescope which was under construction. Hubble wrote:

... The surveys to about the practical limits of existing instruments present as alternatives a curiously small-scale universe or a hitherto unrecognized principle of nature. A definitive choice, based upon observational criteria... may not be possible until results with the 200-inch reflector become available.

This problem, as well as other problems in cosmology, waited for new generations of astronomical instruments and new generations of astronomers.

The best known relation in observational cosmology and the main test is the 'red-shift – stellar magnitude of the brightest galaxies in clusters' relationship (Figure 3). It is known as the Hubble diagram. Deviations of the curve from linearity at large distances reveal information on the nature of the red-shift and on the parameters of the cosmological model. However, it is very difficult to separate the contributions of different effects; unfortunately, it is not possible to do this even now.

As for the nature of the red-shift, which was especially important for Hubble, it can now be considered as solved. Almost all specialists now believe that the red-shift is caused by the expansion of the universe. Ageing of quanta would be different in different spectral ranges and would produce blurring of the images of distant galaxies and other phenomena which are not observed.

Recognition. Facets of his personality

Hubble's Law was recognised practically immediately by nearly every astronomer. This was not surprising. Scientists were all ready: observers had been looking for a relation between velocity and distance to galaxies for many years already and had seen hints of it. The relation was implied by theoretical concepts and, finally, the author of the discovery worked with the largest instrument in the world and was the best expert on extragalactic astronomy. Only Shapley can be said to have doubted the matter at the first stage. Shapley thought that the observational material on radial velocities was not yet adequate for conclusions, that the scatter in galaxy luminosities was too large. Shapley warned, and was later proved to be quite right, that entire stellar clusters might have been mistaken for high-luminosity stars in very distant galaxies used as the most important distance indicators. However, we cannot deny a hypothesis that Shapley's critical attitude reflected a regret about the lost opportunity to have made the discovery himself. Shapley thought that distances and velocities may be related ten years before Hubble's first publication; he even dropped one sentence about it in his paper on globular clusters. Answering Shapley, Hubble wrote:

My paper, will you realize, is merely a preliminary correlation of the data available and makes no claims to finality. In a few years we should have sufficient new data to re-examine the question in a comprehensive manner. I believe that a relation will still be found but whether it will be linear is perhaps an open question.

Soon Shapley could ascertain himself that the red-shift law had been reliably established. Those colleagues of Humason at Mount Wilson who doubted whether he correctly identified spectral lines and did not overestimate the radial velocities of nebulae, had to recognise that their doubts were unfounded.

There can be no doubt that Hubble was flattered to be told that Einstein had a very high opinion of his work. The great physicist had planned to go to the USA for some years already. Finally, in the winter of 1930–1, Einstein, invited by Millikan, arrived at Caltech where Lorentz, Sommerfeld and other outstanding European scientists had lectured before him. Here, in one of the lecture halls of the institute, he listened to the talk by the observers Hubble and Humason and the theoretician Tolman. He came up to the blackboard several times to discuss the implications of the new discoveries. Tolman, who had spent several years in Germany, translated his words into English. Einstein visited the Mount Wilson Observatory and was in raptures. The New York Times on 3 January 1931 reported the following words of Einstein: 'New observations by Hubble and Humason... concerning the red-shift of light in distant nebulae make probable the assumption that the general structure of the Universe is not static'. At one of the meetings with American scientists, Einstein said that Hubble's discovery of the dependence of the red-shift in spectral lines of spirals on the distance to nebulae led to a dramatic concept of the spatial structure of the universe. In all his publications, Einstein invariably referred to Hubble with the greatest respect.

A year later Einstein visited California again. A well-known photograph of Einstein together with Hubble and de Sitter, who in 1932 was lecturing the staff of the observatory on cosmology, was taken on this visit. A year later Hubble met Lemaitre at Mount Wilson; Lemaitre also expressed his respect for Hubble's achievement.

Hubble had been working at Mount Wilson for fifteen years already. These were his most fruitful years. One success followed another. He began with excellent studies of gaseous nebulae in our Galaxy. Then he switched to extragalactic nebulae. From 1924 to 1929, in a mere five years, Hubble achieved outstanding results: he demonstrated that extragalactic nebulae are stellar systems and thus confirmed the theory of the island structure of the universe, he developed a general classification of galactic and extragalactic nebulae, and finally, he discovered the most awe-inspiring and grandiose phenomenon: the expansion of the universe. Five years later, Hubble together with Humason confirmed the red-shift law at greater and greater distances, and studied the distribution of galaxies up to the limit of the capabilities of the 100-inch reflector. Hardly any other astronomer, after Copernicus and Galileo, brought about such a revolutionary transformation of our understanding of the universe, and within a surprisingly short time.

The years of recognition had arrived. Hubble was invited to lecture at American and British universities. He received honours and medals, he became *doctor honoris causa* of several universities.

In spring 1934, Hubble crossed the Atlantic in order to deliver the Halley lecture at Oxford. Foreigners were given this honour in Great Britain only on very rare occasions. For the topic of his talk Hubble chose the red-shift law. Oxford University conferred the degree of *doctor honoris causa* on its former student.

During this visit to the British Isles, Hubble went to a meeting of the Royal Astronomical Society where another American, Harlow Shapley, delivered the Darwin lecture. In his turn, Hubble presented the Bruce medal to the Englishman Alfred Fowler and delivered a speech. However, not only solemn ceremony took place at this meeting, there was also a scientific discussion on the problems of the expanding universe; the theoreticians Milne, Lemaitre and McCrea and the observers Hubble, Shapley and Oort took part.

In autumn 1934 Hubble was invited to Yale University to give a lecture course. Beginning in 1901, such courses had already been organised 24 times on the money donated by the children of Mrs Silliman in memory of their mother. Many outstanding scientists were invited to lecture: J. J. Thompson and Lord Rutherford from England (Nobel Prizes for physics), Walther Nernst from Germany and S. Svante Arrhénius from Sweden (Nobel Prizes for chemistry), T. H. Morgan from the USA (genetics) and W. Bateson from England, who never got the Nobel Prize. Lectures were also delivered by the mathematician Hadamard from France and by others. The only astronomer invited before Hubble was Campbell, who spoke about the motions of stars and especially about measuring their radial velocities.

Now it was Hubble's turn. He had a lot to say to his audience. His lecture course was entitled 'The Realm of the Nebulae'. Two years later, Hubble's lectures were published as a special edition and this book became one of the best astronomical reviews of the middle 1930s. The presentation was clear, there were numerous illustrations – photographs obtained by the 100-inch telescope; but the main asset was what Hubble said. Hubble shared with his audience and now with his astronomer readers the results obtained with the largest instrument in the world, mostly by himself. Two years later, the book was published in Germany, translated by Kippenhahn. It was reprinted in the USA in 1958 and then in 1982, after Hubble's death. Moscow astronomers also greatly valued the book.

Hubble wrote a preface to his book in February 1936. The book emphasised the outstanding role played in astronomy by the 100-inch reflector. At the same time, it demonstrated its limitations. A new, still larger instrument was needed to penetrate space still further.

At just this time, Hale's, Hubble's and other astronomers' dream of the 200-inch telescope was at last coming true. On 10 April 1936, a small train arrived in the afternoon at the Pasadena railway station: a locomotive, two cars and a flat car on which was mounted the glass disc for the future mirror of the telescope. Were it not for the great depression, Corning Glass would have been unlikely to get involved in manufacturing this giant disc. It involved great difficulties. But astronomers were lucky, in those years of recession the company was ready to grab any job promising good returns. For the same reason, Westinghouse Electric agreed to carry out the installation of the instrument. An almost completed tower, a huge building 137 feet in diameter, was waiting for the telescope at Palomar Mountain.

The train moved only in daylight and it took it sixteen days to cover the 3000 miles from the east to the west coast of the country. Thousands of people gathered along the railway in order to have at least a glance from afar at the painstakingly packed disc which was to become the eye of the giant reflector. The disc was carefully transported to the California Institute of Technology for grinding and polishing; astronomers expected that the mirror should be ready by the end of 1941. However, World War II broke out and the mirror was lifted to Palomar Mountain only eleven years later, in November 1947. Hale did not live to see the day, he died in 1938.

Hubble again visited Oxford in autumn 1936 and gave three lectures under the general title 'The Observational Approach to Cosmology'. Hubble again spoke about observable space as a typical sample of the universe, while he was discussing the significance of the red-shift. He was especially interested in what really stood behind the red-shift – the real expansion of the universe or some as yet unknown law of nature which forces light quanta to lose their energy over the millions of years they travel through space. Hubble's lectures made up his second book.

In 1935, Columbia University, on the recommendation of the National Academy of Sciences of the USA, conferred the Barnard Gold Medal on Hubble. This was a rare and valuable distinction conferred once every five years. Hubble and Barnard had something in common: they were both observational astronomers and both obtained their main results by applying photographic techniques.

Two years later, Hubble won a very unexpected prize. On 4 August 1937 he discovered a new comet denoted as 1937 g. It was the seventh appearing in that year and the fifth among the newly discovered ones. The comet was a very undistinguished one, faint and tailless, and was monitored by several observatories. The Astronomical Society of the Pacific conferred on Hubble the Gold Medal, the 165th award for discovering new comets.

In March 1938, Dr H. M. Jeffers, the President of the Astronomical Society of the Pacific, who was to step down that year, presented the Catherine Wolf Bruce Gold Medal to its 33rd laureate. The laureate was Edwin Hubble.

It so happened that until relatively recently the Bruce medal was regarded as the highest award in astronomy. Until the end of the 1960s, the Nobel Prize Committee, following strictly Alfred Nobel's will, did not award the prize to astronomers. The attitude of the committee has changed only in the last twenty years – astronomy was equated in its rights to physics and Hans Bethe, Hannes Alfvén, Martin Ryle and Antony Hewish, Arno Penzias and Robert Wilson, Subramanyan Chandrasekhar and William Fowler obtained this award. If Hubble had lived longer or the committee changed its attitude earlier, his name would definitely appear on this list.

The woman who established the medal, Miss Catherine Bruce, did nothing in science but was undoubtedly an outstanding woman. She was one of the first to receive university education in the USA, she knew Latin, German, French and Italian, she drew professionally and for some reason supported astronomy above everything else. Science was unimaginably poor at that time. Even if it was possible to construct a large telescope, there was usually no money for all the rest, so that a typical observatory would not have enough funds to subscribe, say, to the *Astrophysical Journal*. Miss Bruce sent from 50 to 50 000 dollars to thirty American and European observatories. Her donations to astronomy just in the last ten years of her life (she died in 1900, at the age of 84) came to 175 000 dollars. In 1897, she donated money for a gold medal named after her, to be presented 'for outstanding contribution to astronomy'; a year later, its first winner was the American Simon Newcomb.

The Bruce medal became essentially an international award. After a proposal of E. S. Holden, the first director of the Lick Observatory and the first President of the Pacific Society, the directors of three observatories overseas, the Greenwich, Paris and Cordoba observatories, and of three American observatories, Harvard, Yerkes and Lick, selected each year not more than three candidates for the medal, and the most deserving astronomer was chosen from these three.

Hubble's predecessor in the list of winners was Hertzsprung; his successor was Shapley, whose contributions to astronomy were great and indisputable, as was Hubble's contribution.

The official greeting to the winner was prepared by Harald Babcock. It said:

The point to be noted here is that Dr. Hubble was the first to establish a reliable scale of distances for objects observable with the 100-inch reflector and farther away than the Magellanic Clouds... The volume of space now subject to exploration is a thousand million times as great as that amenable to quantitative discussion only twenty years ago.

Babcock continued:

I shall not attempt a discussion of the cosmological significance of Dr. Hubble's observations. They certainly have reacted strongly upon the thinking of the greatest theoretical physicists and astronomers, have stirred the depths of philosophical speculation, and, perhaps, more important than anything else, have clearly shown the need for still greater instrumental power. It is not an overstatement to say that the convincing data assembled by him with the 100-inch telescope have been a powerful argument for the construction of the 200-inch reflector.

On the evening of 21 March 1938, having received the medal, Hubble delivered a public lecture 'The Nature of the Nebulae'. For some reason, Hubble chose not to speak about the realm of countless remote galaxies nor about their motion that had occupied him in recent years. In a sense, Hubble turned back to the source and spoke about the Andromeda nebula and its satellites, their structure and population, their cores, globular clusters, Cepheids, novae, the 1885 supernova, and about bright non-variable stars. From the standpoint of the history of astronomy, the most interesting point is that already in 1938 Hubble was only a small step away from discovering the doubling of the scale of extragalactic distances. By that time, about 120 novae had been found in the Andromeda galaxy and it had been shown that according to the distance determined by Hubble, these novae are fainter by about one magnitude at the maximum of their brightness than their analogues in our galaxy. Groups of objects of identical nature in these two stellar systems were appreciably different. Hubble was obviously worried. In his lecture he said:

Thus the apparent discrepancies assume unusual importance; they indicate either intrinsic differences between the groups or errors in the interpretation of the available data, and no progress was possible until the question was settled.

Hubble's colleague Walter Baade happened to lead Hubble away from the correct solution. Baade noticed that photographs of the Andromeda nebula were only infrequently taken on large reflectors, at most several times a month. As a result, one usually had to extrapolate somewhat the brightness curves at the maximum, taking into account the rate at which brightness was decreasing at a later stage when the star was quite far from the maximum. However, bright novae in our galaxy were getting fainter very fast right after the maximum and only then was the rate of brightness decrease reduced. If all this were taken into account, the novae in the Andromeda nebula and in our Galaxy could become quite similar.

At that time these arguments looked acceptable and only in 1953–5, when the young Halton Arp got access to the 60-inch reflector and was able to photograph the Andromeda nebula each night, was a large variety of novae revealed. Some of them, quite bright at the maximum, varied very rapidly, while others, much fainter, varied slowly. The problem could not be solved as simply as Baade expected. The relation that Arp had found between the brightness of the novae and the rate at which they become fainter is in fact an altered form of another, well-known dependence: the absolute luminosity of a star is the higher, the shorter the time needed for this star to drop in brightness by three magnitudes. This relation is justly connected with the name of the American astronomer McLaughlin. He had already pointed out in 1939 that the visually observed galactic novae vary faster, on average, than all novae considered together; in 1945, McLaughlin made use of the novae in the Andromeda nebula and reformulated the relationship in its final form, via absolute stellar magnitudes. Actually, Hubble had been the first to notice the relationship, at least in the case of the Andromeda galaxy. When discussing the novae in this nebula, he pointed out in his lecture: 'The brighter novae seem to fade more rapidly than the fainter'. To be absolutely accurate, we have to note that F. Zwicky had somewhat earlier, in 1936, formulated a hypothesis on a relationship between the maximum brightness of novae in the galaxy and the rate of decline, and mentioned some corroborating facts. Neither McLaughlin nor Hubble, nor many others later on, ever quoted Zwicky's work. We do not know what caused this oblivion.

Hubble had known for some time already that the globular clusters in the Andromeda nebula were also fainter than their counterparts in the Galaxy. However, this contradiction did not unduly worry him. He thought that nothing precluded globular clusters from being really different in different galaxies. Clusters in the Magellanic Clouds did look as if this was the case.

It was Baade who succeeded in solving this contradiction fourteen years later, in a very different manner. He found that the distance to the Andromeda nebula is actually considerably larger and that the discrepancy in the luminosities of the objects in the two galaxies is caused only by this factor. Hubble was still active when this important step was made in establishing the new distance scale; however, he never mentioned in any of his articles how close he himself was to the discovery and how he let it slip between his fingers.

Less than a year had passed when the Franklin Institute awarded its Franklin Medal to Hubble, to emphasise Hubble's contribution to science. Then the *Journal of the Franklin Institute* published Hubble's report on the motion of our galaxy among nebulae; it was the lecture that Hubble had read after the ceremony for receiving the medal.

By the middle 1930s, Hubble had become one of the best known astronomers in the USA and the world over. It would not be illogical in this situation if he had plunged into the maelstrom of administrative positions, representative duties and high-level meetings. However, all this was alien to Hubble – he loved science and devoted all his energy to it. Sandage remembered: 'He also did not participate very much in formal society activity as many professional scientists do... Though a member of the American Astronomical Society, he never held office, he rarely went to meetings.... He held himself aloof from the wear and tear of scientific debate. He made no effort to become a politician of astronomy like Harlow Shapley or Otto Struve, like any of the leaders... he just did not participate in that...'

Actually, Sandage was not quite right. Hubble did not refuse to attend the meetings of the society, which usually took place twice a year, but he did it only several times in all his life. Even in these cases, he only appeared in the midst of a large number of colleagues, when the society organised its congresses on the grounds of his observatory or not far from it. Hubble was also elected to the Council of Directors and as the Vice-President of the society. In the 1940s he became a member of the executive committee of the American Astronomical Society, which represented American astronomers at the International Astronomical Union. In all likelihood, he did not accomplish anything important in these capacities.

Sandage says: 'He was a very private man; he did not interact very much with his colleagues'. His correspondence with other scientists is kept in the Hubble archive at the Huntington library. We assume that Hubble did not really care about preserving his letters. Nevertheless, it looks like 'he did not have vast correspondence with many people' (Sandage).

When he did write, these letters were, typically, purely professional

scientific messages to Shapley, Slipher, de Sitter and others, at most
several letters in many years. Mayall was the only one with whom
Hubble kept up a correspondence for 20 years, ever since both of them
started counts of galaxies. Mayall wrote:

To one interested in the same field of research, Hubble was unfailingly helpful and
encouraging, for his letters often ended with 'Good luck, and any way I can help
out, let me know'. In replies to questions or requests for information on specific
nebulae, his answers were invariably explicit and complete, and frequently even
more than one had expected.

Sandage said, possibly recalling three years when he worked together
with Hubble:

He was always kind to young people, very helpful.... He was like a stern father
but not so stern that he was unapproachable.

Nevertheless, as a result of the enormous difference in age (almost
forty years), contacts between these two outstanding astronomers of
our century were mostly professional. Hubble was not surrounded by
students and assistants, in contrast to, say, Shapley. Says Sandage:

He did his own work; he never had assistants till the very end when he was ill;
he worked very hard and his whole life was dedicated to his work.

Hubble as he was seen in his mature years by colleagues at Mount
Wilson was noble in his appearance and demeanour, serious and aloof,
not resembling at all more open and spontaneous Americans; he even
looked haughty, very much immersed in his work.

In fact, Hubble could be very different. Friends who were not professionally connected with the scientist saw a different man, different facets
of his personality, sometimes quite unexpected.

The quiet Woodstock Drive, with abundant trees and shrubs, lies about
three miles to the south of the observatory office at Santa Barbara, in
San Marino township close to Pasadena. Number 1140, a two-storeyed
stone wall house under a tiled roof, half hidden by tree branches and
bushes, can still be seen at the end of the street. It is in this house that
Hubble lived. It is a typical Californian house in which scientists and
professors, intellectuals of modest means settled during that period. The
house is not large by American standards. Sandage, who often visited
this house, remembers that it had five to seven rooms. Hubble's study,
where he did much of his work, was on the ground floor, together with
a Spanish style living room. The house was quite ample for this childless
household. Several years before her death in 1980, Mrs Hubble sold it

to a family with two children. The house was too cramped for the new owners and several years later they had added to it.

Hubble and other astronomers would drive up from the valley to the top of Mount Wilson for observations, typically three or four nights a month. The rest of the astronomer's life and work proceeded in the valley, at the office of the observatory or at home.

Mrs Grace Hubble, who devoted her life to her husband (she was 'a homemaker', in Sandage's words), was the soul of this house. They seemed to be a happy couple. Sandage recalled: 'I did visit often at their home ... and saw how they interacted. They treated each other with really great respect. Mrs Hubble was very devoted to her husband and he was certainly an intellectual in all his discussions with her'. They did not lead a secluded life. The house on Woodstock Drive was visited by various extraordinarily interesting people. Among them, we find Americans and British people whom Hubble had met at Oxford and remained friends with for many years, and people who came to visit the nearby Hollywood. One of the personal friends of the Hubbles was Igor Stravinsky. The Hubbles were very close friends with the family of the well-known English writer Aldous Huxley, the grandson of the biologist Thomas Henry Huxley (a colleague and associate of Charles Darwin). They met in October 1937 when Huxley and his wife settled in California and first came up to Mount Wilson; Hubble showed them the star-studded night sky.

Hubble knew another member of this highly talented line of Huxleys well, Aldous' elder brother, the biologist Julian Huxley. Igor Stravinsky, the Huxleys and writers and actors from Hollywood used to be invited to this house on Woodstock Drive for lunch. Mrs Hubble, who graduated from Stanford University, had a considerable knowledge of literature, art, music and architecture. The archive still holds her correspondence with regular friends and with famous friends, such as, for example, the British film actor George Arliss, who played famous historic personalities and was chosen as the best actor of 1929–30; and Sir Hugh Seymour Walpole, the British novelist, scriptwriter and collector of books and manuscripts. Both Walpole and Arliss travelled widely, and had lived in England and in the USA; fate even had Walpole go to Russia where he served as a paramedic in the Russian Army during World War I and saw the events of 1917 in Petrograd.

Among Hubble's friends and acquaintances were Sir Charles Richard Fairey, at whose house Hubble usually stayed on his visits to England, the Irish playwright and writer Edward John Morton Plankett, 18th

Baron Dunsany, Walt Disney, the creator of the inimitable Disneyland, the British Ambassador to the USA Philip Henry Kerr, 11th Marquess of Lothian. Prime Minister Lloyd George made Kerr his secretary and assistant when preparing the Versailles Treaty. After Hitler occupied Czechoslovakia, Kerr realised that the policy of pacifying the aggressor had to be rejected; he regarded his principal task as ambassador as being to organise the supply of war materials from the USA to Great Britain. Thus the people with whom Hubble and his wife met, made friends and were in correspondence, included a great composer, outstanding writers and actors and numerous other men and women among Californian intellectuals. Hubble was not a man to be lost among this splendid society. First and foremost, he was a scientist, a serious and stern researcher, but he was also very interested in the humanities: history, classical literature, philosophy, and the history and philosophy of science. Hubble was well known in South California beyond his professional circle. In 1938, after Hale had died, Hubble became a member of the Board of Trustees of the library and art gallery founded in San Marino by Henry Huntington, the railway tycoon of the American West. Here, in a world-renowned culture complex, one can see the most rare books and manuscripts, including the first edition of Gutenberg's Bible, a collection of English paintings of the last two centuries, rare tapestry, china and sculptures. Hubble remained on the Board of Trustees until the end of his days; in 1954, the library published in his memory a collection of his public talks, *The Nature of Science and Other Lectures*.

Hubble loved books. In his house, he compiled a rich collection of various publications and books on astronomy. Antique publications were especially valuable. After his death, the observatory obtained, according to his will, more than a hundred books of the 16th, 17th and 18th centuries: the manuscript of Pliny Sr, the *Epitome on Cl. Ptolemaei Magnus compositionem* by Purbach and Regiomontanus, the translation of Apian's *Cosmographia* into Spanish, *Writings of Sacrobosco* with a preface by Melanchthon, a humanist and theoretician of Lutheran Protestantism, the second edition of Copernicus' treatise, books by Galileo, Kepler, Hevelius, Riccioli, Newton's *Principia...* These books, which sometimes changed owners many times over several centuries, are all marked as Hubble's posthumous donation and are all kept in the library of the observatory.

Civil duty

In June 1939, the well-known European and American astronomers gathered for the last time before World War II in Pasadena in order to discuss the structure and dynamics of galaxies. Among them were Lindblad and Oort from Europe, and Mayall of the Lick Observatory and a number of Mount Wilson and Caltech scientists represented the USA. The conference was not formal, so that the hosts and their visitors could talk in detail about the problems of interest, in a straightforward manner.

By that time, Hubble had achieved considerable progress in studying the shapes of galaxies. Working with the 60- and 100-inch reflectors, he had photographed 800 bright galaxies of the Shapley–Ames catalogue and about 1500 fainter objects scattered over the entire sky. This was quite sufficient to return, almost a quarter of a century later, to the classification of galaxies. Even though there was a long way to go in completing the analysis and drawing conclusions, and although Hubble never finished this project, it was clear that the value of the existing classification system was confirmed and that the entire scheme appeared to be logical, elegant and continuous. The situation with dynamic studies was not as satisfying: there were not yet enough spectral data.

The participants at this conference did not leave us their recollections, so that we do not know what they felt when parting. Nevertheless, it is certain that they could not be sure about the time of their next meeting: they realised that the situation in Europe was getting increasingly grim.

American astronomers got together once again, soon afterwards. The 62nd meeting of the American Astronomical Society was convened at the beginning of August at California University. This last pre-war meeting assembled an especially large number of participants. American astronomers did continue to meet in subsequent years but many of them were occupied with quite non-scientific problems and could not come at all.

At this meeting, Hubble presented a short communication on barred spirals, pointing to a resemblance between them and normal galaxies. They did not differ from ordinary galaxies in any physical characteristics except the connecting bar. It was, then, logical to suspect that the bar could be formed in response to external, not internal forces. However, Hubble failed to find any causal relationship between the orientation of bars and the neighbouring galaxies. The problem of the participation of the bar in the rotation of a galaxy remained absolutely unclear.

The meeting elected Hubble as the vice-president of the AAS for
1939–41.

Many participants at the congress visited the observatories on the
Pacific coast: Lick, Mount Wilson and Mount Palomar. In Pasadena,
they could see the 200-inch mirror. Those who worked at the Yerkes
Observatory were likely to be especially impressed by the fact that the
size of the Cassegrain hole in the giant mirror was exactly equal to
the size of the objective of the famous 40-inch refractor. The visitors
could see at Palomar Mountain the completed tower of the new telescope
where it was being installed, and the tower of the future 48-inch Schmidt
camera. On one of the nights, the visitors observed the sky through the
60- and 100-inch telescopes at Mount Wilson. An exception was made
for these visitors, since the large reflector was typically absolutely off
bounds for visitors. They climbed the stairs to the observation platform
where they were met by Hubble who demonstrated to them, in the
Newtonian focus, Mars and the familiar planetary nebula 'Saturn' in
Aquarius. Astronomers are quite familiar with the 'visitor effect'. The
guests were unlucky: images were rather poor and this time they could
not see anything on the disc of Mars.

Hubble was working intensely. Together with Baade, he was studying
interesting objects that Shapley had just discovered. Shapley wrote:

We were taken by surprise in 1938 when Harvard plates unexpectedly, and almost
accidentally, yielded two sidereal specimens of an entirely new type. The gamut
of galaxies had already been run. All the forms had long been fully described.
There were spirals, spheroidals, irregulars, with many variations on the spiral
theme. The newly found organizations in Sculptor and Fornax did not seem
essential in order to fill in a natural sequence; they were not logically necessary.
On the contrary, they introduced some doubt into the picture we had sketched
– they suggested that we may be farther than we think from understanding the
world of galaxies.

On one of the photographs, Shapley found a swarm of about 2000
small dots scattered over an area of less than one degree of arc along
one side. This swarm in Sculptor resembled surprisingly well some
sort of defect on the plate. Test photographs demonstrated, however,
that the dots were real celestial objects. Was it a cluster of very distant, very faint galaxies or of stars? Large telescopes were needed
to investigate these new objects, so Baade and Hubble exposed a series of plates using the 100-inch reflector. The distribution of the
stars (the members of the Sculptor system were shown to be stars)
is strikingly uniform. The system has no core, no star clusters, no

supergiant stars or diffuse nebulae. The system was very dissimilar from NGC 6822, the galaxy that Hubble had analysed in detail many years before. Forty variables were identified in the Sculptor system. Two of them were likely candidates for Cepheids. This result implied that the system lay at a distance of 84 000 parsec, far beyond our Galaxy. This was a faint dwarf galaxy of about 1000 parsec in diameter.

The Fornax system, larger and more than twice as distant, was found to be very similar to the Sculptor system. The Local Group of galaxies expanded to two new members with unusual characteristics.

Hubble, together with Baade, started another large programme: photographing the sky along the Milky Way at low galactic attitudes. The purpose was to outline better the boundaries of the galactic zone of avoidance and to probe for areas where it would be possible to penetrate as far as possible and study very thoroughly the structure and population of our galactic system. The plan was very extensive: it was necessary to photograph fields of stars in six bands every 5 degrees of latitude, over 100 degrees in longitude. The desired data were indeed partially collected but no publication ever appeared. Transparent windows were found in the Milky Way band. One of them, the so-called Baade window in Saggitarius, made it possible to reach the galactic central region.

In 1939 Hubble tackled two exciting problems: estimation of the mass and the direction of rotation of galaxies. As in the case of binary stars, one can find the masses of pairs of nearby galaxies moving around their common centre of mass. The radial velocities of a number of binary galaxies had already been measured at that time. Hubble carefully analysed Humason's material but realised that the errors in the radial velocities were too large. He was only able to estimate the upper limit: 10^{10} solar masses. Together with Humason, Hubble started a programme of more exact measurements. The Mount Wilson report of 1940–1 informs us that the relative motions had already been found for 20 pairs of galaxies; the velocities were rather small both in compact and in loose binary systems. It was an indication that the average mass of galaxies could not exceed 10^{10} solar masses.

We do not know whether Hubble grew indifferent to this topic or decided to put it aside for a time; at any rate, any reference to it disappeared from the pages of the reports. Hubble never resumed this work, and never published a line about it.

It is very likely that Hubble's thoughts were at that time already occupied with the menacing events taking place in Europe; out of the programme that he outlined for himself, he was able to complete only the study of the direction of rotation of the arms in spiral galaxies.

The problem may have looked simple, at least in principle: for a galaxy seen almost from the side, one has to measure radial velocities at its ends and to determine what side of it leans towards the observer. The problem of the direction of spiral arms is not a detail of secondary importance, it is related to a more general problem of what spirals are, of how and why they evolve. It is natural to assume that the arms trail as a result of rotation, that is, the arms lag behind the core, the farther away they are from the rotating central part. However, the theory which Lindblad had been developing since 1926 predicted the opposite: spiral arms are leading the way, they are moving with the concave side forward. A band of dark dust matter could indicate the orientation of a galaxy in space but such galaxies are rare (Hubble was able to find only three such objects in a thousand at the beginning of his survey). Later their number was increased to 15. Hubble showed that they all rotate in the same direction, that is, they either all twist or all untwist. Only four galaxies enabled Hubble to choose among these two possibilities. The absorption band in these galaxies was projected very conveniently onto the bright central part and thus determined the position of the galactic plane with respect to the observer. The answer was unambiguous: these spiral galaxies had their arms twisted backwards. One could hypothesise, then, that this property is common for all spirals.

Hubble, together with Mayall, first reported these results at the annual session of the National Academy of Science in Washington, in April 1941. However, considerable time passed before Hubble prepared a detailed paper. It was received at the *Astrophysical Journal* in January 1943, when its author had already left the observatory. It so happened that this paper was to be the last but one scientific, rather than popular or review, paper of Hubble's. Lying ahead were the war years, the second war in Hubble's life.

On January 1933, President Hindenburg of Germany called Adolf Hitler, the head of the 'National-Socialist German Workers Party' to form the new government. The German Nazis came to power, immediately unleashing total suppression of democratic freedom and bloody terror, implementing their anti-human racial theory and rabid ideas of world domination. Freedom-loving intellectuals who refused to live under Hitler's new order were thrown into concentration camps. Those whom the Nazis failed to arrest left the country one after the other. Albert Einstein also crossed the ocean.

The year of 1933 marked the beginning of preparations for a new world war. Hitler dreamed of implementing the programme presented in his *Mein Kampf*. In March 1936 the German army occupied the demilitarised Rhineland, and in summer it fought on Franco's side against Republican Spain. In spring 1938, Hitler annexed Austria without a shot being fired. The Munich agreement with England and France freed Germany's hands for new acts of aggression. Germany took over the Sudetenland from Czechoslovakia and then Czechoslovakia ceased to exist as a sovereign state.

Europe was on the brink of a new war which would inevitably pull into its orbit tens and hundreds of millions of people.

On 31 August 1939 a group of SS men organised a provocation at the Polish–German border, and on 1 September, Hitler sent German armies into Poland, under a pretext. That was the beginning of World War II. On 3 September, England and France declared war against Germany.

Hubble had spent his happy years of youth in England. He loved the country, had many friends there, and used to visit there from the USA. A friend of Mrs Hubble remembered that on one of their walks along Sussex hills Hubble picked up a piece of flint from the ground and said that he would carry it to California and put it on his desk to remind him permanently of Britain. Everything taking place in Europe and concerning England was always close to his heart, especially so in those troubled times.

The first months of the war on the western front were a 'phony war'. The French army, the British expeditionary corps and the German army faced one another on the two sides of the powerful Maginot and Siegfried lines but all was quiet at the front: no battles were yet raging across Europe.

Military activities did take place in the Atlantic, where Nazi surface raiders and submarines were hunting British vessels, trying to disrupt supply routes. The hunters kept clear of American ships, but already on the day of the declaration of war a German submarine had sunk a British liner with a number of American citizens on board.

Nevertheless, England was not yet seriously shaken by war. Life went on as it used to before. As usual, the Royal Astronomical Society was having its sessions. On 9 February 1940 the society president Professor Plummer announced that Hubble had been awarded the Gold Medal, the highest sign of merit given by the Royal Astronomical Society. Typically, the winner of the award gives a speech during the ceremony, outlining his work. This time, Hubble's trip to England would be impossible. Plummer had to describe for himself the merits of the new laureate, emphasising that his excellent work continued the contributions of the preceding investigators of nebulae; among these, Plummer mentioned quite a few Englishmen: the Herschels father and son, Lord Ross and Isaac Roberts.

Addressing Mr Schoenfeld, who was officially representing the American Embassy, the President solemnly said:

This medal represents the highest mark of our appreciation for individual work contributing to the progress of the science to which we are devoted. I now place it in your hands and beg that you will be so good as to transmit it to your distinguished countryman, Dr Edwin Hubble, of the Mount Wilson Observatory. With it go our heartiest congratulations and our fervent hope that a long and fruitful life still lies before him.

In spring 1940, Hitler's army occupied Denmark and Norway. Early in the morning on 10 May, the German armies moved forward on the front from the North Sea to the Maginot line, invading the neutral Netherlands, Belgium and Luxemburg, and then rushed into France. On 22 June, France signed the act of capitulation in the Compienne forest, in the same railway car in which the Armistice had been signed after World War I.

When World War II broke out, the United States government announced its neutrality. However, the fall of France forced a new attitude on the world situation, and made the threat to the USA obvious. On the day when Hitler's armies started their onslaught, President F. D. Roosevelt addressed the following words to the Eighth Panamerican Congress of Scientists:

... This very day, the tenth of May, 1940, three more independent nations have been cruelly invaded by force of arms. I am glad that we Americans of the three Americas are shocked and angered by the tragic news that has come to us from Belgium and the Netherlands and Luxemburg... Until now we permitted ourselves by common consent to search for truth, to teach truth as we see it... Can we continue our peaceful construction if all the other continents embrace by preference or by compulsion a wholly different principle of life? No, I think not... I believe that by overwhelming majorities of all the Americans you and I, in the long run if it be necessary, will act together to protect and defend by every means at our command, our science, our culture, our American freedom and our civilization.

Hubble, with the analytical approach of a first-class scientist, understood political situations quite well. He realised at a very early stage that the policy of pacifying the aggressor cannot lead to anything acceptable.

When Germany committed its first act of aggression, Hubble said with absolute conviction: 'This is a World War and we are in it.' Hubble was not on the side of President Roosevelt and his line of internal politics. Sandage said:

He was very conservative in his politics. He was very anti-liberal... that's fairly common with people in his position... He did not approve of the liberal policies of Roosevelt. That was quite common in the 1930s. He, like so many conservatives at the time, saw the social order being changed – from the conservative point of view – and most people at Mount Wilson were conservative.

As far as we can judge, Hubble did not approve of the President's international politics either. However, Roosevelt's words on the day the Germans invaded Belgium, the Netherlands and Luxemburg could hardly raise Hubble's objections. He could not stay aloof quietly going on with his research, and chose to head the Southern California Joint Fight for Freedom Committee.

In October 1940, Hubble gave a public speech calling for immediate help to Britain. The German army was on the coast of the English Channel. Germany started an all-out battle for supremacy in the air; when they failed to win this battle, they shifted the direction of their air strikes to London and other cities. In Oxford, so dear to Hubble, whole blocks of houses lay in ruins.

Hubble had a gift of persuasion. This time he wanted especially strongly to make it clear to everyone that the position not only of England was critical but of America as well, since the ocean was not a very reliable defence line. Addressing his audience, he said:

We have watched the explosion of total war in Europe. We have seen the little nations, striving to maintain the decencies of civilization, suddenly blotted out by the stamp of an iron heel. We have seen the shambles of Rotterdam. We may turn to the source material, and, in the pages of *Mein Kampf*, learn what Hitler and Hitlerism really mean. We may turn to Hitler's conversations, to learn what he plans for us.

It was clear to Hubble that America was not yet ready for war, that it did not have a sufficient number of planes, ships and weapons. He continued:

So, in this emergency, we lack the one essential factor – time, which waits for no man. And we have one shield to shelter us while we prepare. That is the armed force of Britain.

Hence, what was needed was to expand the military industry, to refuse to resort to any political manoeuvring, to put off social reforms in the country, to enforce strict discipline and to give all possible aid to England, to its people. Hubble concluded his speech with these words:

It is this people who are defending us with their blood, toil, tears and sweat. Let us say to each other,   Here and here did England help us,   How can we help England.

In November, Hitler's air force shifted its strikes from cities to large industrial centres and ports of England. Intense combat was going on at sea, and the losses of the English fleet were rising steadily. It seemed that the Germans would be able to implement their 'Sea Lion' plan of invading the British Isles; the plan was prepared in detail at the German military headquarters. Hubble had this to say:

... But suppose that Britain fell, and fell soon. If that catastrophe should happen then we would be looking askance at one another and asking how we have used the time that was given us... If we compare production costs with national income, England and Germany are devoting more than half their productive efforts to the war. This means between 4 and 5 hours each day. Our effort, in our race against time, amounts to about an hour, if we include our aid to Britain. And we persist in the illusion that we can have both guns and butter in our struggle for security; we talk as though social gains were paramount when the survival of our democracy is at stake.   That, gentlemen, was the picture as it appeared then, and as it appears today. A declaration of war is the most direct and efficient method of meeting the situation, and it is long overdue. There is a job to be done, and we must get on with it.

Hubble understood full well that the defeat of England would have catastrophic consequences for the USA as well:

A Nazi dominated Europe, even if it were restricted to the area west of Russia and north of the Mediterranean, certainly a modest estimate, would still be two thirds as large as the United States, and would contain 400 million inhabitants. The industrial capacity is greater than our own, and it is largely self-sufficient in raw materials such as iron and coal.

Germany would be able to control the seas, one could expect some countries of the American continent to join Germany so that the enemy would be at the doorstep of the USA.

To conclude, Hubble said:

We all want peace. But it must be peace with honor. Peace at any price is a religion of slaves. The freedom we cherish is a heritage from brave men who the world over, since time began, have fought to establish and maintain it. If there is one lesson that History has taught us, it is this, strong men can determine their own destiny.

Hubble had formulated these phrases – 'peace at any price' and 'peace with honour' – some time earlier. Arthur Compton, Nobel Prize winner, later the director of the Manhattan atomic project, had asked Hubble in spring 1940, on behalf of the American Association of Scientists, to put his signature under their peace address. Hubble firmly refused, stating that he could not sign a document containing no defined position. He wrote in his reply to Compton:

I like to believe that the scientist, outside his peculiar activities (which are concerned with the world as it is, not as it ought to be), is better equipped than most men to clearly see the implications of various lines of action, and to recognize the enormous values of integrity, of tolerance and of freedom. And, consequently, I think he should be the first to recognize the importance of fighting for those ideals when they are seriously threatened.

Perhaps we should first strive to firmly re-establish those ideals in our own political administration. It is not unlikely that, had we commanded the respect of the world for our national integrity, and had we openly supported England last year, that support might have prevented the very wars which your resolution deplores.

Our ideals, it seems to me, should combine high principles with the courage and the strength to maintain them. Peace at any price is a religion of slaves. I will not sign a resolution into which it is possible to read such a defeatist attitude.

In the spring of 1941, the entire Balkan region was in the hands of German Nazis and Italian fascists. Practically all continental Europe, except for the European part of the USSR, was under Hitler's heel. On 22 June Hitler invaded the USSR. Soviet troops, suffering enormous losses, retreating eastward, were trying to stop the enemy on the vast territory from the Black Sea to the Arctic Ocean. The defence of the Brest citadel, the battle of Smolensk, the defence of Leningrad, Odessa, Tallinn, Sevastopol, the desperate fight for Moscow – these were events watched closely by the entire world. We do not know what Hubble knew about them. There is no doubt, however, that the war in the East could not stay 'unknown' for him. It was clear to everybody that only the USSR and Great Britain were standing against Hitler, that the main burden and the decisive role in the war against Hitlerism lay with Russia.

On Armistice day, 11 November, Hubble made a speech to American war veterans. He again spoke about Hitler's annexation plans, about the need for Americans to realise at last the danger of what was going on for America itself, that America's isolationism must be discarded. He again appealed to the USA to declare war immediately against Hitler's Germany. His arguments were not new but they were clear-cut and impressive. Here are a few sentences from his speech:

I am not telling you that we should fight England's war, or Russia's war. I am

telling you that this is *our* war, that our miscalled aid to Britain is aid to ourselves
and never was anything else ...

'We know what war is like. This nation was created by war. Whenever our
national existence is in danger, we must, and we will, fight again. That time is
come – it is now, today...

If an American Expeditionary Force is necessary to insure the destruction of
the Nazis, then an A.E.F. must go abroad. It won't be a matter of choice, it will
be a grim necessity.

Hubble delivered his speech a mere six weeks before the Pearl Harbor
disaster. On 26 November 1941 an aircraft carrier armada of the Japanese
fleet set sail for Hawaii, in the dead of night, with radio transmitters
silent. On the morning of 7 December the first wave of Japanese bombers
went into action and bombed the main naval base of the USA Pacific
Fleet at Pearl Harbor. The second wave followed.

Nowadays only an unusual memorial over one of the American ships
on the bottom of the harbour reminds one of the night which took the
lives of almost two and a half thousand American naval crew.

On the next day, after President Roosevelt's address to both Houses,
the Congress of the USA declared war on Japan. The war spilled from
Europe to the vast areas of the Pacific and the contiguous territories.

Hubble followed his own code of honour. His actions were logical and
he realised that patriotic speeches were no longer sufficient. Immediately
after Pearl Harbor, Hubble made an unsuccessful attempt to enlist in the
army. He, a former major, was not accepted into the infantry. Quite
suddenly, Hubble was invited to the artillery command and informed
that they had intended, for some time already, to use his services at the
research centre of the Aberdeen Proving Ground in Maryland.

Hubble left the observatory at the beginning of August 1942 and went
to the east coast of the country. The huge Aberdeen proving grounds
were spread out on the coast of Chesapeake Bay. This was the main
centre of military research, that had been established during World War I.
Here artillery, tanks and other weapons were tested.

In summer and autumn 1942 important developments occurred in the
course of the war: the Anglo-American allies landed in North Africa.
It was not yet the second front, so anticipated in Europe. In October,
English troops started their offensive at El Alamein. In the Pacific, a
battle began between the naval forces of the USA and Japan in the
Solomon Islands. But the main developments happened on the Soviet–
German front. Forcing the Soviet troops to retreat, the Germans reached
the Caucasus Mountains. Simultaneously, they were advancing towards

the Volga. The USSR was in a precarious position. The Stalingrad battle began and was to rage for many months. We do not know what Hubble thought about this battle. We can only recall the words of the President of the USA. In May 1944, evaluating the role played by the battle on the Volga banks, F. D. Roosevelt acclaimed the defenders of Stalingrad, saying that their glorious victory had stemmed the tide of aggression and constituted the turning point in the war waged by the allied forces against the aggressor.

Many bloody battles occurred later. Soviet, American and British soldiers continued to die but the course of the war had already turned.

In 1946, Hubble was reminiscing at the Sunset club about his work at the Aberdeen proving grounds:

... the centre of the Aberdeen establishment was a Ballistic Research Laboratory which had a small staff in peacetime and was suddenly called upon to carry an enormous load of work. There was a thoroughly competent regular army officer, Colonel Simon, acting as director, and the outstanding staff member, Robert Kent, who is the equal of any ballistician in the world... He went with the army, a student rather than an administrator, and in spite of the obstacles, became the recognized leader in the country...

So, when the need for expansion could be foreseen, there was no reservoir from which to recruit. All that could be done was to call on research men from other fields in the hope that they might 'mug up' the new subject somewhat faster than laymen. Simon and Kent called for advice from scientists in the east, mostly National Academy men, and made a list of possible victims against the day of need. My name, they told me, was top of the list because ballistics has a curious affinity with astronomy, and, moreover, as a line officer in the last war I might appreciate the significance of some of the problems as viewed from positions in front of the guns as well as behind the guns.

Hubble very rarely left home without his wife; at that time he wrote one letter after another to faraway California from the Atlantic coast. Hubble described how he reached the place of his new job, how he settled down, how he joined the work. Sometime later, Mrs Hubble joined him at the proving grounds.

The war was going on but the country was not ready for it in the field of ballistics. Hubble said:

We start each war with the weapons used at the end of the previous war. In between, when we should concentrate on research and development, we just can't be bothered. I asked the generals why, and they murmured 'appreciations'... at any rate, the army seems to feel they can't justify research during peace, and, of course, they haven't the time during war.

Gradually, Hubble mastered the new job. He wrote in a letter to his wife at the beginning of September:

I am more and more impressed with the place. It is not the home of genius but it knows the answers to many problems and how to get the answers for others.

Much had to be done. Hubble's division was to be responsible for providing all the firing and bombing tables. They had to calculate tables for each type of gun, each type of shell and bomb. They had to generalise the basic data on new projectiles, and do it fast, and to be concerned with bomb sites and with fire-monitoring equipment.

A small group could not cope with such a large job. It had to be expanded to the size of an institute. Hubble had not had a large number of people under his command for many years. At the observatory, the people working alongside him had been half a dozen first-class experts who needed no control. Now he had to recall the time when he was young and, as Major Hubble, commanded a battallion.

The division of exterior ballistics grew to 280 people: officers, enlisted men, civilians and WACs. The department was divided into a theory group, laboratory units and field units. Hubble said that WACs were a godsend because they were selected from college graduates in mathematics and physical sciences and came at a time when it was extremely difficult to find civilians, especially with their sort of training. They produced several dozen tables a month, all required by the army and air force.

Unexpectedly, the theory group was engaged in activity connected with the USSR. On 2 June 1944, 'Flying Fortresses' landed for the first time at the landing strips of the Poltava air force centre in Ukraine. This started the shuttling operations of the American air force. The planes took off from territories liberated by the American army, bombed important targets in Rumania and Hungary, which at the time were Hitler's allies, and then landed in Russia. Here they loaded Soviet-made bombs and unloaded them onto the enemy on the way back. Soviet airfields serviced more than a thousand Flying Fortresses, and thousands of tons of Soviet bombs were dumped on the enemy. The bombing tables were computed by WACs of Hubble's department. Hubble said:

One real feat was to produce bombing tables for Russian bombs without any data except qualitative descriptions. These tables were for the use of our bombers on their return trips when they started landing in Russian territory.

Very much work had to be done by theoreticians who for the first time worked on the general theory of projectile motion influenced by all possible forces: gravity and drag.

The laboratory units carried out experimental investigations of the behaviour of small models of bombs and shells in an indoor firing range and in a supersonic wind tunnel. The characteristics were tested in flight with life-size models photographed in two planes.

Hubble's subordinates were also working very hard on the firing ranges. A projectile had to be photographed in flight, by instruments recording its coordinates and time. It was especially important to study the flight of rockets launched from a plane. The behaviour of a rocket after it separates from a plane, and the behaviour of a plane after it has launched the rocket, were studied with high-speed cameras superposing time marks on the record. This work made use of purely astronomical techniques, so Dirk Rheil, the astrometrist of the MacCormic observatory, headed one of the research groups. His name was much later given to a crater on Mars. His group developed efficient methods of measurements on photographic plates; these methods served ballistics very well until the advent of radar and other novel equipment.

Hubble's days were very full. He organised the work of all divisions, designed ballistic instruments, and was dead tired every day, having walked miles and miles over the territory of the proving grounds. He had neither time nor strength for other occupations. Hubble's articles on red-shift problems, meant for the general public and not very different in content, still appeared in popular magazines in 1942 and 1943. They contained no new information. In 1944, Hubble could not write a line on science. It is strange, but the lives of two very different men in different times and places may sometimes follow very parallel routes. Hubble, the leader in observational cosmology, had to switch during World War II to the same problems that the cosmology theorist Alexander Friedmann had had to engage in during World War I – bombing control and bombing sights.

The observatory, despite the many thousands of kilometres separating it from the war front, already felt the burden of war in 1940. Its staff members, especially optics experts, began to work on defence projects. After Pearl Harbor, more and more of them dropped whatever they were doing in pure science. People left for military projects, two young men from the technical assistance group volunteered to serve in the army. Visitors from Europe came for the last time in the summer and autumn of 1941 – Erik Holmberg arrived from Sweden and later two Turkish officers were sent down from the State Department. Links connecting Mount Wilson with European astronomers were getting thinner and weaker and then broke down almost completely. Only a few publications reached the observatory from overseas by roundabout ways.

War required increasing efforts of the country. Many astronomers followed Hubble: Ralph Wilson who was a specialist on stellar motions, spectroscropists William Christie, Theodor Dunham, Olin Wilson, Gustaf Strömberg, and others. Those who stayed also switched to defence work. The last but one winter of the war was especially difficult. The heaviest snowfall in the history of the observatory struck it in February 1944. Sixty inches of snow fell in three days. The observatory was blocked for two and a half months, and the services were too short of people to clear the mountain road. Repairs of the electric power lines had to be put off for quite a while.

The fortieth anniversary of the installation of the first telescope at Mount Wilson was not celebrated in any way. The 60-inch reflector was idle: there was no one to work with it. Only solar instruments were operating, and Humason and Baade were still working on the largest reflector. Humason kept measuring radial velocities of galaxies and accumulated more than 400 by the end of the war. During the first season after victory, the number reached a round figure of 500.

Walter Baade was born in Germany, received his astronomical education there and moved to the USA in 1931, as a member of Hubble's department. He thought of becoming an American citizen only just before the war, had prepared all the necessary documents but lost them someday when moving house from one place to another. He did not bother to resume the procedures. The war surprised him as a citizen of an enemy country. There was a threat of internment, but he managed to get permission to stay at the observatory. There was no question of using Baade for defence assignments, so he spent all his time observing on the 100-inch reflector. The Los Angeles valley looked very unusual in those years. The authorities were afraid of Japanese attacks, so blackout rules were enforced stringently everywhere in the huge city and its satellites. It was on such nights that Baade was able to conduct one of his most important studies. Baade turned to the Andromeda nebula. He recalled that

by that time Hubble had left this field, which he would undoubtedly have cultivated more if he had not in the meantime become absorbed by the red-shift and the expansion of the universe, and he now turned entirely to the cosmological problem.

Baade noticed on one of the excellent photographic plates unmistakable signs that the amorphous central part of the nebula close to a spiral arm looked resolvable into individual stars. In autumn 1942, Baade started to prepare in earnest for the final resolution of stars in the central part of the Andromeda nebula. Taking all possible precautions, Baade was able to resolve the central part of the nebula into stars in August and September of the next year, and later resolved its companions, the NGC 205 and M32 galaxies. When Mayall told him that the NGC 185 galaxy to the north of the Andromeda nebula had almost the same radial velocity and may be connected with Andromeda, Baade resolved this object into stars. as well. The neighbour of NGC 185, the NGC 147 galaxy, was also split into individual stars. All the objects investigated contained numerous red stars, considerably brighter than the red stars in open clusters of our galaxy. Baade realised that he had found in other galaxies the same stellar population as in globular clusters. This is how the concept of two types of stellar population was born: young population I, already known to Hubble, and older population II.

The discovery was published in 1944. The moment the September issue of *Astrophysical Journal* reached the USSR (the journey took half a year with one of the Arctic convoys of the Allies), the head of the Moscow stellar astronomy group, Pavel Parenago, immediately realised the significance of this achievement. He published a translation of Baade's entire communication in the first post-victory issue of the *Soviet Astronomical Journal*; he regarded it as the main achievement of American astronomy during the war years.

Baade made his outstanding discovery during nights with especially quiet atmosphere, when stars appeared as dimensionless dots in the eyepiece of the telescope. However, Baade was not losing time even when the nights were not so good. He began to photograph the Andromeda nebula systematically, hoping to find emission from gaseous nebulae, that is, HII regions which are bright in the red hydrogen line. Hubble had tried to solve this problem earlier, but failed. Baade remarked that Hubble was 'a victim of the fact that he could at that time observe only in the blue; red plates of any sensitivity were not available at that time'. In Baade's 'red' negatives, the Andromeda nebula was virtually spattered with gas clouds.

Baade also made considerable progress in another problem that interested Hubble. When scanning the plates of the Andromeda nebula, he discovered many new globular clusters, usually fainter than those Hubble had seen before him.

In all likelihood, Hubble in his time was close to resolving the northern companion of the Andromeda nebula, the NGC 205 nebula, into separate stars. Hubble wrote in his book *The Realm of the Nebulae*: 'Very faint stars are more numerous than would be expected for foreground stars alone, and some of them may be associated with the nebula'. Baade found numerous faint variables in this companion and also in the southern companion of the Andromeda nebula, M32, during the war years. More than a quarter of a century has passed since Baade's death, but astronomers still lack any understanding of what these variables are. Baade disliked bringing his work to the level of published papers; in fact, we even know of his greatest achievements, such as the discovery of ionised gas clouds and globular clusters in the Andromeda nebula, from publications written by other authors who were able to work with his material.

The year 1945 was approaching. Much had changed in the world by that date. Almost the entire territory of the USSR and most European countries were freed of the enemy yoke and the war had moved onto the land of Hitler's Germany. The USSR and its allies were so powerful that the ultimate annihilation of the enemy was close at hand. Following the successes of the Red Army, the situation became favourable for the offensive of the Allied armies. By the end of March, the armies of the USA, Britain, Canada and France had reached the Rhine along its whole length, and Soviet and American soldiers met on 25 April on the River Elbe.

The final battle took place around and inside Berlin. Hitler was repeating over and over again that 'Bolshevism will bleed to death at the walls of the German Empire... Berlin will remain German'. However, it was no longer possible to prevent history running its course. The red banner of victory was flying over the Reichstag early on 1 May and 24 hours later the Berlin garrison capitulated. On the night of 8 May 1945, the Russian Marshal Georgi Zhukov, the American General Carl Spaatz, the Air Force Marshal of Great Britain Arthur W. Tedder and the French General Jean Lattre de Tassigny accepted the unconditional surrender of all the German armies, navy and airforce.

We do not know how Hubble spent the Victory Day. He never discussed it with Alan Sandage, his youngest and last pupil. Older colleagues of Hubble left no memories. We only know a photograph taken on 8 May 1945, in which Hubble is seen standing by the supersonic set-up of the Aberdeen Proving Ground.

Hubble did his patriotic duty with distinction. He could also be satisfied with how his work was regarded: in 1946 he was awarded the Medal for Merit, created for civilians in the USA and its ally countries for outstanding contribution to the war effort.

The citation accompanying the award said:

Dr. Edwin P. Hubble, for exceptionally meritorious conduct in the performance of outstanding service as Chief of the External Ballistics Branch of the Ballistic Research Laboratory, Aberdeen Proving Ground, Maryland.

Through his leadership and untiring effort, and by ingenious adaptation of his high scientific training in other fields, Dr. Hubble directed a large volume of research in exterior ballistics, which increased the effective fire power of bombs and projectiles.

This work was facilitated by his personal development of several items of equipment for the instrumentation used in exterior ballistics, the most outstanding development being the high-speed clock camera, which made possible the study of the characteristics of bombs and low velocity projectiles in flight. The results of such studies have contributed greatly to the improvement in design, performance, and military effectiveness of bombs and rockets.

This was a very high distinction. An identical medal with the gold eagle – the symbol of the USA – against the background of a dark blue ring with white stars was awarded in the same year to Enrico Fermi, Robert Oppenheimer, Harold Urey and other physicists who were engaged in the race to overtake Germany in developing atomic weapons.

The war was now over both for the soldiers and for those who had worked for this victory in the plants and in military laboratories, very often far from the places where they had lived before. Astronomers, one by one, began to gather at Mount Wilson. By December 1945, Hubble had also returned to the observatory.

Hubble got to know a great deal during his work at the proving ground. He saw the tremendous destructive power of modern weapons – non-atomic weapons but nevertheless weapons much more menacing than those used in World War I. Hubble himself was working on making the weapons even more effective. The war was over and it was time to think of the consequences of further increases in military power. Many honest American scientists, atomic scientists who were frightened by the result of their effort, and Hubble after a very different job, were facing this dramatic question.

Hubble returned to Mount Wilson firmly believing that war had no more place in the life of mankind. In his public speech 'The War That Must Not Happen' delivered in Los Angeles in 1946, he said:

Scientists in their explorations have recently entered new fields, and in these have found knowledge of a quite new kind. Through technology, this knowledge is being used in a way that will convert war into suicide... The last war ended with jet propulsion, long-range rockets, guided missiles and atomic bombs – the devastating combination. You have heard of them all, and, in addition, you have doubtless heard some references to the possibilities of bacteria and radioactivity...

Warfare with the new weapons will be the ruin of civilization as we know it. If you accept this thesis, and are interested in survival, you must accept the conclusion, namely, the absolute necessity of eliminating warfare... The world today has become so small, so to speak, in the accessibility of all its regions, that it is no longer possible for any nation to achieve safety in isolation. We are part of an organic whole, members of one body. Even if against our wishes, we must cooperate successfully in order to survive. There will be no defence from now on against attack. War and suicide – we must accept these as synonymous terms, for they amount to that.

It is difficult to avoid an impression that these words must have been said today, when all living creatures on the Earth are threatened by an even more real menace; Hubble said them almost half a century before us, at the very beginning of the atomic age.

In Hubble's opinion, mankind could survive only if a world government with powerful punitive forces were created. Although his approach was usually very penetrating, Hubble was obviously unable to fathom to what extent the political map of the world would change and that plans for maintaining security, which were discussed in the West not only by Hubble, would be unacceptable to governments.

Hopes crushed

Back at the observatory Hubble immediately resumed his work. He had, first of all, to continue the project interrupted by war: to reconsider the classification of galaxies. Hubble again scanned hundreds of photographs of galaxies obtained during thirty years with the 60- and 100-inch reflectors, identified their common features, singled out attributes of each type. Hubble wanted to prepare an 'Atlas of Galaxies', in order to demonstrate clearly typical samples of each subdivision of his classification. This work was not completed. Only his remarks on the negative plates and fragments of the manuscript were prepared. Sandage, who knew Hubble's plans, created such an atlas in 1961. Astronomers refer to this atlas as the Hubble Atlas of Galaxies. Hubble scrutinised especially those galaxies which were resolvable into stars. His purpose was to study high-luminosity stars with variable and permanent brightness and then use them as individual distance indicators. He discovered variables in the galaxies M51 and M101 beyond the bounds of the Local Group, and then discovered novae in M81.

Hubble was able to measure the magnitudes of the brightest stars in about 80 spiral and irregular galaxies. Their average magnitude in the

Virgo cluster was about 20. Hubble discovered an interesting property of this population of galaxies. He found that the luminosities of the brightest stars increase with increasing luminosity of their parent galaxies. This meant that one cannot always assume their luminosities to be identical when using these stars as distance indicators; at the same time, with the difference between their luminosities and the luminosities of galaxies known, one can, and must, introduce appropriate corrections.

Each year, Hubble discovered more and more supernovae. First he found them in 1946 in the NGC 3977 and NGC 4632 galaxies. Then in March 1947, he found a supernova in NGC 3177. Baade plotted the light curve for it, and Humason studied its spectrum. In the spring of 1948, Hubble found another supernova in the NGC 4699 galaxy.

To discover four supernovae in two years is an excellent result, especially when the search was not intentional. However, Hubble never published a word about them. The supernovae were mentioned only in the reports of the observatory, from which the new data travelled to the general catalogues.

We cannot say that Hubble was not interested in supernovae. He had studied the remarkable supernova Z Centauri in the galaxy NGC 5253 in 1922, jointly with Lundmark. True, the profound differences between an ordinary nova and a supernova were not yet known at the time, so the authors of the publication simply referred to it as a nova. This nova had occurred rather long ago, in 1895, and Hubble and Lundmark compiled the data and plotted the light curve. Z Centauri was the brightest supernova after the famous 1885 supernova in the Andromeda nebula, reaching a magnitude of 7^m (we would add now, the brightest except for the 1987 supernova in the Large Magellanic Cloud). The supernova was hundreds of times brighter than the parent galaxy, which consisted of hundreds of millions of stars. Actually, the NGC 5253 galaxy is not a large one.

Analysing the plates made with the 100-inch telescope, Hubble and Lundmark found a faint star close to where the supernova occurred. However, it would be wrong to assume that it was the formerly bright supernova. This would imply that the proper motion of this star would be too great. Hence, they concluded that no object is observable at the supernova location, and the range of its brightness variations exceeded 13.5^m.

In 1923, Hubble scanned the plates of the well-known galaxy M87 in which the Pulkovo astronomer I. A. Balanovsky had discovered a supernova in 1919. A year after it occurred, it was still observable on

Mount Wilson plates; it became fainter than the detection threshold only a year later. Hubble's observations again demonstrated that the amplitude of supernovae was definitely greater than 10^m. Hubble made two remarks in his paper about the Balanovsky star and about the M87 galaxy itself. Curtis had already written in 1918 that a sort of straight ray extends from the core of this galaxy. Hubble was probably the first to notice that it consisted of five starlike condensations. This was the famous jet – an ejection whose nature was deciphered in 1955 by I. S. Shklovsky. Another interesting feature was a multitude of objects of magnitude about 20, surrounding M87. Hubble had discovered globular clusters in a distant galaxy, almost a decade before this was done for the nearby Andromeda nebula. However, he wrongly believed at the time that these were not clusters but stars.

In 1928, Hubble organised a systematic search for supernovae in the cluster of galaxies in Virgo, where Wolf had already identified five supernovae over 17 years. Hubble, and later Baade, started observations with the 10-inch camera which allowed them to photograph the entire cluster on a single plate. Later Glenn Moor joined them and in seven years of observations, they were able to discover the first supernova in the NGC 4273 galaxy. The star was fainter than the entire galaxy by only two units of magnitude. Its spectrum was also obtained. The supernova of NGC 4273 is the only case for which Hubble published a paper describing his discovery.

In 1938, comparing Duncan's old plates with recent ones, Hubble discovered two more supernovae in the galaxies NGC 4038 and NGC 3184.

Evidently, Hubble was still interested in supernovae in the later period too, and delivered a public lecture about these spectacular objects, but he stopped any active efforts in this field. Zwicky, from Caltech, then started a vigorous search for supernovae. First, he tried to do it (unsuccessfully) with a very small (only 4-inch) camera but then continued his search with the 18-inch Schmidt telescope at Mount Palomar. He discovered his first supernova within a year, and had brought their number to 17 by 1940. This was only the beginning of numerous discoveries that Zwicky and his assistants made in subsequent years.

In 1946, many changes took place at the observatory. Adams, who had been in charge of the observatory for more than 20 years, resigned from his post. He was replaced by Professor Ira Bowen of Caltech; Bowen was not an observational astronomer but rather a physicist who applied his knowledge of spectroscopy to the study of gaseous nebulae.

A change of leadership is always a complicated process, especially in a small team. No-one knows how relations between people will evolve, what new impulses will be provided by the new director...

The first peaceful year brought important and happy news. The California Institute of Technology and the Carnegie Institution agreed that the Mount Wilson and Mount Palomar observatories must be merged. Work was resumed on constructing the 200-inch telescope, which had been interrupted in the war years. The 48-inch Schmidt camera was soon to start operation. The work of the joint observatory headed by Bowen began on 1 April 1948. This was the creation of an unprecedented concentration of the largest instruments at one place and in the hands of the most gifted researchers, such as Baade, Humason, Zwicky and Minkowski of Hubble's department, and others.

Hubble had worried for some time already about who must steer the progress of science, and how, especially fundamental science. He had sent a letter at the end of the war with his deliberations on the subject to Vannevar Bush, who was then heading the Carnegie Institution. As we can conclude, the Institution had probably formed a wrong idea of the scientific role of the observatory, and the style of management even in the observatory itself irritated Hubble because it was getting more and more bureaucratic. The doubtful competence of functionaries who insisted on their way of running science was very irritating to him. His annoyance can be seen in what he wrote to Bush:

The administrator represents the Observatory to the Institution and to the public, while the astronomers must be content with the private satisfaction of their work and the recognition of their colleagues...

As for the management of the astronomical observatory, Hubble's ideas were given in the same letter:

The broad planning of research should be placed in the hands of a committee of senior astronomers headed by a chairman or leader who has the deepest and broadest available background, knowledge and vision in the field of astronomy. The chairman should represent the staff, insofar as research is concerned, to the Institution and to the world of science...
The research committee, let me add, should have the assistance of a small group of consultants from allied fields. Such men would be invaluable in suggesting new types of instrumentation, techniques, and methods of interpreting data, and in keeping the astronomical research group fully informed concerning the development and implications of advances in physics, theoretical physics and mathematical theory which touch the problems of astronomy.

Hubble ended his letter with the following words:

Insofar as pure science is concerned, administration should assist research and should not direct research, that direction of research in the limited field of the Observatory (namely, that part of astronomy which should be studied with large telescopes) should be left to leaders in the field. On the other hand, the leaders should be freed from as much as practical of all affairs that do not directly pertain to research and research programs. I am fully convinced that a research leader, plus a competent executive officer, is the proper solution of the problem, and I am confident that the consensus of opinion among the senior astronomers of the country will support this view.

All scientists will agree even now that the goal is to minimise bureaucratic domination, to rely on the opinion of competent experts who know best of all in what direction to move and how to conduct fundamental research.

It is difficult to find out whether Hubble's letter ever played any role, but a committee for developing long-range research programmes was created immediately after the two observatories were merged. This was an important task. The 200-inch telescope was suitable for working on most diverse research problems but scientists could not let themselves be distracted by secondary aspects; life dictated that they formulate the main objective that could not be solved by any other instrument: the study of the general structure of the world. The observatory was represented in this committee by Baade, Merryll and Nicholson and Caltech was represented by Tolman and Oppenheimer, who had left the position of director of the nuclear research centre at Los Alamos. The committee was headed by Hubble.

However, Hubble was engaged not only in planning the future of astronomy. There was a need to develop recommendations on financial management and other financial matters, on coordinating and merging various ancillary services of Mount Wilson and Mount Palomar; in addition, he had to think about the astronomer training programme at Caltech. Hubble, who was also appointed to the joint management committee together with two other representatives of the observatory, its former and current directors, had to divert his attention to these time-consuming occupations as well.

American astronomers, and with them the entire world astronomical community, waited impatiently for the beginning of a new stage in the history of their science: the start of operation of the giant 200-inch reflector. Hubble's anticipation of this event may have been greater than anybody else's.

By the end of the 1940s, Hubble was definitely the leading figure in world astronomy. The great discoveries that he had made with the 100-inch reflector were recognised by all astronomers. He became virtually a classic figure in science while still alive. New honours were showered on him after the war, in addition to those he had won before it. In England, he was elected an Honorary Fellow of Queen's College, he became a corresponding member of the French Institute and an honorary member of the Academy of Science in Vienna, Austria. In the University of California, the faculty remembered his law degree and made him a *doctor honoris causa* in law.

Hubble was never elected a foreign member of the USSR Academy of Sciences. Perhaps, someone was frightened in the years before the war by the corollaries of the red-shift. Alas, astronomers were never in luck in the USSR. In 1924, when the German Max Wolf and the Americans Hale and Campbell were elected, Hubble was just starting on his glorious path. This was followed by a long 'dark age', a period of dead silence, and only in 1958 did the Swedish physicist and astronomer Hans Alfvén become a member of the USSR Academy.

Hubble's name became familiar not only to scientists but to the general public as well. In 1948, his portrait appeared on the front page of *Time* magazine. Hubble was the first astronomer among the celebrities that the journal presented to its readers. Only later, many years after this issue, did the magazine print the portraits of Maarten Schmidt and Carl Sagan.

Hubble participated in a series of lectures on the radio together with other American scientists. Hubble described how mankind had successively discovered for itself the space around our Earth: first the Solar System, then the world of stars, and finally, the space beyond our galaxy, containing uncountable multitudes of similar huge stellar systems. Hubble pointed especially to two basic features of the surveyed part of the universe: uniformity, on average, of the distribution of galaxies in space, even though they form groups and even huge clusters; and the solidly established fact of the red-shift, which was most probably evidence for the expansion of the universe.

The talks of Hubble, Bowen, Shapley and Russell appeared in 1945 in a volume called *The Scientists Speak*. The main message of all these talks was this: the study of the universe must continue to greater distances. The hope of success was connected to a large degree with the 200-inch telescope whose construction had just been completed.

In April 1947, Hubble delivered the Morrison public lecture on 'The 200-inch telescope and some problems it may solve' in Pasadena. There were quite a few such problems, but Hubble selected three of them. Not only the general public, but also astronomers wanted an answer to the question of whether or not thin mysterious channels exist on Mars. The new telescope would be instrumental in solving this problem. It would collect as much light as a million human eyes, so it would be possible to photograph the Martian surface with very short exposures. It would then be possible to select from numerous negatives the best photographs made on nights with an especially clear and quiet atmosphere.

Another very important field for astronomy and astrophysics is to obtain detailed spectra of celestial bodies and to study their chemical composition. The giant telescope was to say its new word in this field as well.

However, the main objective of the new instrument was to solve cosmological problems. Hubble was able to establish that the red-shift is proportional to distance up to the limits accessible to the 100-inch reflector. Inevitably, one had to go forward and check whether the relation is maintained at greater distances, whether space continues to be uniformly filled with galaxies. The red-shift results in decreasing the apparent brightness of galaxies. However, with the recession velocity in the previously observable part of the universe not exceeding a tenth of the velocity of light, this decrease stayed below 10%. With the new giant reflector, the region of the universe accessible to observation was to increase twofold, so that the anticipated weakening of the farthermost galaxies would reach 40–50 percent, thus becoming easily observable.

Hubble wrote:

We may predict with confidence that the 200-inch will tell us whether the red-shifts must be accepted as evidence of a rapidly expanding universe, or attributed to some new principle of nature. Whatever the answer may be, the result will be welcome as another major contribution to the exploration of the universe.

The general public may even have formed an impression that the new telescope was meant exclusively to solve the cosmological problem. This response generated jealousy in the rest of the staff of the observatory. Its spectroscopy experts once called a press conference in order to describe their work. Nobody told Hubble about this meeting, so that he was accidentally informed about it by a reporter he knew. He came quite unexpectedly into the library where journalists were listening patiently to descriptions of very important but not very exciting spectral investigations. Hubble was also asked to say a few words. As a result, his description of the origin of the universe, and of the outstanding role of Humason's spectral studies was so colourful and engaging that the reporters forgot about everything else and only Hubble's words appeared on the pages of papers and magazines.

Hubble was full of hope. In a year or two, he would start working on the solution of the grandiose problem. He was sure, not without foundation, that after Hale's death he had become *de facto* master of the new telescope. However, things worked out differently.

At the beginning of the 1970s, the American Institute of Physics started an extremely interesting project: it was decided to put on magnetic tape the recollections of numerous scientists. Documents never store all the details, and sometimes important decisions are formulated during unofficial meetings at which scientists freely exchange their opinions. They work out the strategy of the search, some people and their ideas move forward, others recede into the background; the true springs and levers of events in science, unknown beyond the narrow circle of participants of such informal meetings, are revealed only at such gatherings.

The historians of science recorded interviews of a hundred American astronomers. Committed to paper, they make several volumes of most interesting memoirs, 10 000 pages in total. Some of this material is still classified: the time to open all the details to the general public has not yet come. However, most of these records are accessible to historians. Something about the last years in Hubble's life also became clear.

On one quiet day in 1948, Hubble, Baade, Bowen, Tolman and several other people got together in a private house in Pasadena. The topic was how to organise work on the 200-inch telescope. Hubble expected that he would have the lion's share of observation time and that he would be able, on moonless nights, to measure stellar magnitudes of faint galaxies and record their counts, as Tolman and he had outlined even before the war. However, other participants of the meeting stood their ground: Hubble's plan in the form he presented it could not contribute substantially to solving the problem of the expansion of the universe. Carefully, trying not to make it painful, but at the same time quite decisively, they told Hubble that his plan had to be rejected. He heard out his colleagues with dignity and quietly, at least on the surface, as should behove the gentleman that Hubble always was. However, he may have been crushed when he realised that all his hopes were dashed. This was not a tragedy on the scale of the entire science, since no one doubted the importance of the problem of the expansion of the universe; the disagreement lay in how to obtain the solution. However, this was a tremendous personal drama for the man who had devoted twenty years of his life to this undertaking.

The participants of this meeting kept no minutes, so the whole event might have remained unknown. All parties to that discussion, except one, are now dead, and they have departed without dropping a hint of how one of the most important decisions of modern astronomy was taken. Only Martin Schwarzschild, the youngest among those who opposed Hubble, depicted to historians many years later how it had all happened.

Meanwhile, the installation of the telescope was nearing completion. The mirror from Pasadena was transported to the tower in the second half of November 1947. Astronomers looked through the eyepiece lenses of the 200-inch telescope for the first time on 21 December.

Half a year later, the instrument was officially opened. Celebrations began on 1 June 1948 when the invited guests got together in the tower of the new telescope, which had been turned into a sort of congress hall for the occasion. First the hosts described the work of the optical and mechanical systems of the telescope. Then Walter Baade outlined the programme of research on the new giant:

After the first great attack on the structure of the Universe, which was started by Hubble in 1925 and in the course of the next twenty years was carried forward by him toward the cosmological problem, we see very clearly where we stand today and what we shall have to do next. What we shall have to do next can be expressed in a very simple sentence: we shall have to strengthen and widen the base on which the future structure of extragalactic research must rest.

This meant that one had to begin with quite a few things: to carry on the photoelectric expansion of the scale of magnitudes of the faintest stars in a number of areas in the sky and thus make it reliable, to continue searching for luminous stars and especially Cepheids in distant galaxies and, using them, determine distances of up to 10 million light years. Only then, having created a stable foundation, was one able to return to the determination of the Hubble constant. Galactic astronomy was facing impressive objectives: it was necessary to determine more accurately the zero-point of the period–luminosity relation for Cepheids. To do this, Baade suggested the study of globular clusters in our galaxy. Using the largest telescopes, it is possible to study faint dwarf stars whose luminosity is well known from surveys of similar objects in the neighbourhood of the Sun. It is then possible to determine reliably the luminosity of Cepheids in clusters. Furthermore, the luminosities of other members of clusters would also become known: variables of the type of RR Lyrae. Baade expected that the Palomar instrument would detect them in the Andromeda nebula as well. These data would provide another check on the luminosity of Cepheids. This would supply astronomers with a truly reliable cosmic yardstick.

On 2 June the managers of the Carnegie Institution, California Institute of Technology and the Rockefeller Foundation visited Mount Wilson in order to have a look at the highly distinguished 100-inch which was already losing its number one position. However, observing through the 100-inch telescope was impossible: clouds came and raindrops started to fall.

On the next night, the distinguished guests observed at the Coude focus of the new, 200-inch telescope Saturn, the famous M3 globular cluster and the cluster of galaxies in Corona Borealis. Both Saturn and the globular clusters were immediately forgotten once the guests realised that the light from the galaxies they observed through the eyepiece lenses, which until now could only be photographed, took 120 million years to reach the Earth.

The ceremonies culminated on 3 June. A. Dubridge, the President of Caltech, solemnly announced: 'On May 10, 1948, the Board of Trustees of the California Institute of Technology unanimously adopted the following resolution which I herewith announce for the first time: "The Board of Trustees of the California Institute of Technology hereby resolve that the 200-inch telescope of the Palomar Mountain Observatory shall hereafter be known as the Hale Telescope"'.

Hubble did not give a speech at the opening of the telescope. Lists of those present (there were about 800 guests) have never been published so it is not known whether he was present at all.

The festivities ended and defects in the main mirror were soon discovered. It had to be dismantled, the aluminium reflecting layer removed, and the mirror repolished. It was accepted as satisfactory only in the autumn. Astronomical tests of the telescope were getting closer.

The year 1949 had arrived. The weather improved after an annoying wait of more than a week. And in the evening of 26 January Hubble aimed the telescope at the familiar cometary nebula NGC 2261 with a variable star, R Monocerotis; it was the very nebula that he had studied at the dawn of his life in astronomy. The first exposure with the new instrument took 15 minutes, and the number marked on the edge of the developed negative plate was PH–1–H, for Palomar, Hale telescope, negative No.1, observer Hubble.

The atmosphere was not quiet the following night, and it was possible to test the penetration depth of the new telescope. Hubble photographed

Kapteyn area No. 57. Here, in a small patch of the sky, was one of the then most reliable photometric standards: a group of stars measured down to magnitude 21. Even though the mirror still required retouching at the turned-up edge and the aluminium coating was somewhat dusty, the result exceeded the most optimistic expectations. During a mere 5–6 minutes, the new telescope gave the same result as its Mount Wilson predecessor during the maximum exposure. This signified that with a dark sky and with a very quiet atmosphere, it was possible in one hour to photograph stars of a magnitude fainter by 1.5^m, that is, fainter by a factor of four, than those accessible with the older instrument.

Another outstanding characteristic was that a plate taken with the 200-inch revealed many more galaxies than stars. As expected, the faintest of them were twice as far away as those detectable with the 100-inch telescope. A wide field of possibilities had been opened up for penetration into the depths of extragalactic space.

Photographic plates of individual galaxies, M87, NGC 5204 and NGC 3359, were exceptionally good. The first of them is a huge elliptic galaxy surrounded by an 'atmosphere' of what was then interpreted as 'supergiant stars', which in fact were globular clusters. They were only suspected on the plates of the 100-inch instrument, while here they were obvious at the very first glance.

A photograph of a late spiral in Ursa Major, NGC 5204, disclosed the presence of a considerable number of bright stars, which could be studied individually.

A well-developed barred spiral NGC 3359 proved to be extremely interesting, revealing on the plate numerous previously unknown details.

The title Hubble gave to his report on the astronomical testing of the new tool was 'First Photographs with the 200-inch Hale Telescope'. However, his article in the popular science journal *Scientific American*, had a title that was much closer to the truth: 'Five Historic Photographs from Palomar'.

After three months, the new telescope had already yielded about 60 plates, and by July of the next year the number of photographs exceeded 500.

The 48-inch Schmidt camera had also begun operations at the beginning of 1949. Now the 200-inch telescope with a very small field of view (of a fraction of the lunar disc) was complemented by an instrument which allowed astronomers to photograph large areas of the sky. This was a powerful combination of telescopes for extending survey and analysis far deeper into space. Hubble, Baade and young Zwicky worked with the new camera. Together with Sandage, the new astronomer at the observatory, Hubble studied the distribution of very distant faint galaxies on several plates.

In July, Hubble left for Colorado. He loved spending his holidays outdoors, but fishing was his passionate hobby. There is an excellent photograph of a contented Hubble in windbreaker and high boots, with a fishing rod in his hands.

Suddenly, this powerful man suffered a severe miocardial infarction. The illness took him out of observatory work for many months to come. His health began to improve very gradually. On 23 October, Aldous Huxley wrote to his son Matthew:

We saw the Hubbles yesterday. Edwin is sufficiently recovered now to be able to spend a little time in his office and to walk as much as a mile or so. He confidently hopes to be allowed to go to Palomar as soon as the mirror is in place after its re-grinding and re-silvering, which have kept it grounded for the last few months.

At the beginning of December, Aldous wrote this about Hubble to his brother Julian:

He got through by the skin of his teeth and is only now beginning to be up and around. Whether he will be able to go on observing at high altitudes is still uncertain. It will be a great blow to him if he can't make use of the two-hundred-inch telescope, which is finally in perfect working... The dome is unheated, the temperature at six thousand feet is often arctic, sleeping habits are badly interfered with. It may be that now his dream of twenty years has actually come true, poor Edwin will not be allowed to take advantage of the new opportunities.

By Christmas Hubble's health had improved further, and together with his wife, who was by now completely worn out by all the worries, he dared to leave Pasadena and spend Christmas with the Huxleys.

Hubble did manage to work with the new telescope after all. Sandage recalled that before his heart attack, Hubble had spent only three observation rounds at the 200-inch telescope, and was very eager to go on. Mrs Hubble was fairly unhappy because of his insistence on going to Palomar, and she accompanied him. It seemed that his health had recovered very well. Sandage could not notice a difference between his pre-infarction and post-infarction states. He resumed work, but he did not observe very much at Palomar. He only had two or three observation sessions after his illness. He slightly reduced the intensity of his activities, most probably because of the insistence of his wife.

Huxley's letters at the end of 1949 enable us to recognise Hubble's other worries at the time, on top of his illness and the uncertainty about a continuation of his work. We find that he thought, among other things, about the fate of scientists in the USSR. Science in the USSR was then very troubled. In 1948, T. Lysenko organised a shameful session of the Agricultural Academy and virtually annihilated biology in the country. Other discussions were going on, in which obscurantists, covering themselves by pseudo-philosophical and pseudo-patriotic phraseology, were ousting true scientists. Hubble knew about that, as did all his colleagues around him.

We do not think that Hubble felt close to astronomers in the USSR. In all his life, Hubble cited Soviet authors only twice: B. P. Gerasimovich and I. A. Balanovsky. This is natural for at least one reason: he was working in a field in which colleagues in the USSR could not compete with him. However, Hubble suspected that the witch-hunt launched in other sciences could finally reach Soviet astronomers.

During this time, Julian Huxley published a book about 'Lysenkoism' (J. Huxley, 1949, *Soviet Genetics and World Science: Lysenko and the Meaning of Heredity*). Aldous Huxley wrote to his brother:

I have been reading your Lysenko book with the greatest interest. What a dismal picture emerges! And apparently the trend is not confined to genetics. Edwin Hubble tells me that there is now a party line in astronomy – one theory of the origin of the Solar System being orthodox and all the rest not.

In another letter, to his son, Aldous Huxley quoted Hubble's remark about those who did not support this orthodox theory: 'They will soon, no doubt, suffer the fate of the bourgeois-idealist Mendelists–Morganists in the field of genetics'.

The situation in astronomy was quite menacing, against the background of what the authorities were doing to literature, the arts, biology and other sciences. For quite some time already, energetic 'science-accompanying' operators had been pontificating to experts on what in astronomy is materialistic and what is idealistic. In 1947, one of the highest-ranking party leaders, A. A. Zhdanov, at a conference on various aspects of philosophy, attacked those who were interested in the problems of the world as a whole: '... Many of Einstein's followers, extending the results of studying the laws of motion in a finite bounded part of the Universe to the infinite Universe, dare to speak about the finiteness of the world, about it being finite in time and space'. Zhdanov then sarcastically remarked that 'the astronomer Milne has even "calculated" that the world has been created 2 billion years ago'. Pronounced by the all-powerful 'expert' in sciences and arts, this sounded as a very menacing warning. Repeating his words, many a henchman started to attack Western scientists right and left, for their 'concoctions', 'idealism and superficialness', for the 'degradation of theoretical thought exposed to idealism'. They mentioned the names of Eddington, Hoyle and others. Hubble's name was not among these, possibly because it is difficult to reject observations, and because Hubble did not pursue problems that were too general, even though the discovery of the red-shift law was his.

Several months elapsed, and then Hubble was able to return to the main work of his life.

Work had begun on the programme outlined for the 200-inch reflector. Hubble and Humason exposed 13 plates during the first four months and discovered seven novae in the M81 galaxy. The weather was especially bad in the winter and spring of 1950–1, so observations were rare. Nevertheless, using his own plates and those obtained by Humason, Baade and Minkowski, Hubble discovered a dozen irregular variables and a considerable number of Cepheids in the NGC 2403, M81 and M101 galaxies. On the first photograph of M101 of this season, on 3 February 1951, he found a very bright star, presumably a supernova. The next year, Mount Palomar astronomers discovered 20 novae and 25 variables in the group of galaxies containing M81. Humason, Baum and Sandage each discovered a new nova, and Hubble discovered the other novae and variables. A dozen Cepheids were definitely identified among the variables. Now it was necessary to wait for the development of faint photometric standards, in order to plot their light curves. Hubble also found the first novae and variables in the galaxies in the Ursa Major and Virgo clusters. An important component of the general cosmology programme of the 200-inch telescope was a search for such distance indicators as novae and variable stars in distant galaxies; this was carried out successfully under Hubble's supervision.

Four years had elapsed since the regular work with the new telescope had begun, and much had already been achieved in the cosmological problem.

Humason was recording the spectra of increasingly distant and faint galaxies. The greatest measured red-shift had already reached 60 940 km/s. By Hubble's estimate, the clusters to which these galaxies belonged were at a distance of 300–360 million light years from us. Using the then unpublished photometric data of Whitford, Hubble found out that, within the accuracy achieved, the red-shift law still remained linear even for such tremendous linear velocities and distances.

Even though the work was far from completed, it was found that two clusters of galaxies in the southern hemisphere, on the side of the sky opposite to those already studied, also obeyed the established red-shift law. Hence, the value of the red-shift was independent of direction. Deviations from the already known relation between radial velocities and stellar magnitudes of galaxies in clusters were also small. This signified that the space between galaxy clusters was sufficiently empty, that it contained no dust which would dim the light from very distant objects.

In 1950–1, Hubble and Sandage started a study of bright variable stars in the nearest galaxies; the Andromeda and Triangulum nebulae. Hubble had discovered them about a quarter of a century previously and now it had become possible to monitor their brightness over a long time. We know now that these are the most massive stars, whose evolution must be especially fast, transforming hydrogen in their central regions into helium. They could also serve, together with other extremely bright objects such as globular clusters, novae and non-variable stars of high luminosity, as good distance indicators for distant galaxies. The last step would be to establish their luminosity using nearby galaxies containing well-known Cepheids.

Hubble added three new such variables to the two that Duncan and Wolf had discovered in the Triangulum nebula. Another bright variable was mentioned by Baade but he never published the discovery; the star was on the edge of the exposed plates and Hubble was unable to study it. In the Andromeda nebula, Hubble knew about only one high-luminosity variable. These were very rare objects, there being only very few of them in galaxies populated with billions of stars.

Very rich observational material was accumulated over fifty years at the Lick, Mount Wilson and other observatories. The only unrecoverable gap occurred in the war years when work almost stopped at Mount Wilson.

The brightness of stars varied in an unpredictable manner: in some years they looked bright, in others they became considerably fainter. Two stars in Triangulum, which Hubble denoted by the letters A and B, were of special interest. The first of them had been getting progressively brighter ever since the end of the last century. However, it suddenly and sharply faded when it had seemed that nothing could stop the growth in its brightness. This star has been observable only through the largest telescopes for almost four decades now. It has transformed into a red supergiant. Shall we witness in the foreseeable future a supernova flaring up in Triangulum, as the theory predicts? Another star, B, was increasing its luminosity not monotonically but through oscillations, increasing in brightness after each cycle. Hubble was able to record its spectrum in one such epoch, in 1940. This was the first case in which it had been possible to obtain, with sufficient details, the spectrum of an individual star in a galaxy removed to a distance of 3 million light years.

Everything pointed to the discovery of a new, previously unknown type of variable star, which deserved the name of Hubble–Sandage object. The title of the paper written by Hubble and Sandage referred to the first of these objects. High-luminosity variables of this sort were also found in M81, NGC 2403, M101 and other galaxies, and new publications were to follow.

At the end of June 1953, the authors sent their paper to the journal, and it was published in November, several weeks after Hubble had died. For Hubble, this was his last publication, and for Sandage, carrying on the work of his teacher, it was his first serious research paper.

In 1953, Hubble was invited to Britain to give the Darwin lecture which commemorated George Darwin, the well-known British astronomer, son of the great Charles Darwin.

On 8 May Hubble read the lecture 'The Law of Red-Shifts' at the meeting of the British Royal Astronomical Society.

I propose to discuss the law of red-shifts – the correlation between distances of nebulae and displacements in their spectra. It is one of the two known characteristics of the sample of the universe that can be explored and it seems to concern the behaviour of the universe as a whole. For this reason it is important that the law be formulated as an empirical relation between observed data out to the limits of the greatest telescope. Then, as precision increases, the array of possible interpretations permitted by uncertainties in the observations will be correspondingly reduced. Ultimately, when a definite formulation has been achieved, free from systematic errors and with reasonably small probable errors, the number of competing interpretations will be reduced to a minimum. The path toward such definitive formulation is now clear and the investigations are under way at Mount Wilson and Palomar.

Hubble outlined for his listeners the history of how the red-shift law had been established: stage one had been its discovery, stage two, confirmation, when Humason was able to obtain the spectra of galaxies receding at a velocity up to 40 000 km/s, working at the limit of possibilities of the 100-inch telescope. Now was the time for stage three – the 200-inch telescope had been created and an impressive cosmological programme had been started.

There can be no doubt that Hubble realised the importance of the revision of each step in establishing the law of red-shifts. As early

as 1951, while delivering a public lecture at Pasadena, he presented
the programme of cosmological research not as his personal project
but as a unified programme of the entire joint Mount Wilson and
Mount Palomar observatory. Now he was able to describe to his British
colleagues how far American astronomers had advanced towards its
solution.

Hubble reminded his listeners, with satisfaction, that already in the
observational season of 1950–1, Humason had been able to measure the
radial velocities of galaxies of 50 000, 54 000 and, finally, 61 000 km/s.
The greatest recession velocity was found for a galaxy in the Hydra
cluster. This was a barrier which had proved insurmountable for the
100-inch telescope. The most important part of the programme of the
new instrument was to revise the distance scale. It had been suspected for
some time already that the distance scale had weak spots. For instance,
the relative brightness of Cepheids and globular clusters and of Cepheids
and novae in the Andromeda nebula were found to differ from those
in our galaxy. The problem was solved by Baade. He established that
Cepheids must be brighter than had been assumed. In that case, all
distances in the universe, which somehow or other were all based on the
Cepheid method, had to be doubled. This also doubled the age of the
universe, so it had to be of at least several billion years. This re-evaluation
removed the contradiction with the geological age of terrestrial rocks:
the Earth ceased to appear older than the rest of the universe.

Observations at Mount Palomar often used photoelectron multipliers,
which were new for astronomers; they had been developed during World
War II. They measured stellar magnitudes of standard stars and the
integral magnitudes of distant galaxies. These quantities were needed
to check the red-shift law when stellar distance indicators proved to be
too faint, inaccessible even to the new instrument. The magnitudes of a
number of galaxies were finally measured in 11 clusters. Hubble managed
to confirm that the linear law of red-shift still held.

It can be stated with some confidence that a definitive formulation of the law
of red-shift will be available before long, in the form of a relation between
red-shifts and apparent magnitudes, out to red-shifts of the order of 0.25... The
implications of the law established over this distance may be traced out to nearly
double that distance by effect on the apparent distribution of nebulae in depth.
Thus, if red-shifts do measure the expansion of the universe, we may be able to
gather reliable information over a quarter of its history since expansion began,
and some information over nearly a half of the history.

These were the nearest perspectives as Hubble saw them. Subsequent steps seemed much more problematic to him. This work would need even more powerful telescopes, which would be extremely costly. Hubble, who knew from his own experience the kind of money absorbed by modern weapons, realised that if the cost of, say, one aircraft carrier were paid out 'to the consolation of philosophy', it would immediately solve all the financial problems of astronomers. Even though no such money was forthcoming, he considered the future of his science optimistically. Hubble concluded his lecture with the following words:

From our home of the Earth we look out into the distances and strive to imagine the sort of world into which we are born. Today we have reached far out into space. Our immediate neighbourhood we know rather intimately. But with increasing distance our knowledge fades, and fades rapidly, until at the last dim horizon we search among ghostly errors of observations for landmarks that are scarcely more substantial. The search will continue. The urge is older than history. It is not satisfied and it will not be suppressed.

Hubble delivered the Cormac lecture to the Royal Society of Scotland, delivered a report to the Royal Institute of Great Britain, was a guest at Greenwich where the Duke of Edinburgh, the Queen's husband, opened the Octagon Room, the first building of the old observatory turned into a museum; he then left for Paris and participated in a meeting of the French Institute of which he had been a member for some time. Then he returned to Bossington House near Stockbridge, the house of his friend Sir Richard Fairey where, as always, he enjoyed fishing. Hubble's colleagues found it amusing that even business trips to England coincided, for some reason, with the season when the trout bite best.

Returning home, Hubble could feel glad with his trip to Europe, having met his old friends and received general recognition.

There was nothing unusual when the day of 28 September 1953 began. All morning Hubble worked in his study at Barbara Street and left for lunch at home. Mrs Hubble met him in a car. They drove along the streets of San Marino, talking about science. Who knows, Hubble may have been thinking then about the expected four nights of observation at the Palomar giant; he was unfortunately not doing much observing too often those days. The irremediable event happened when the car was already slowing down at the house – it was a stroke. Edwin Hubble died only about three weeks before his 64th birthday.

The next day, American papers informed the public about Hubble's death. The obituary in the *New York Times*, the same paper that many years earlier had announced his discovery of Cepheids in the Andromeda nebula, briefly outlined his life and his scientific achievement. It also pointed out another, perhaps little known, facet of Hubble's activities. Hubble had realised full well the environmental menace of the pollution of air by the notorious smog; for several months he had been head of the Pure Air Council of Southern California.

Nevertheless, when one reads the paper of that day, one tends to feel disbelief and rejection. Not because something incorrect was written about Hubble, but because the words said about Hubble are placed alongside the obituary, and even the portrait, of the former Nazi Hans Fritsche, Hitler's favourite radio commentator, the mouthpiece of the regime that the famous scientist had fought relentlessly. A strange aspect of the freedom of the press...

Hubble's colleagues – Humason, Adams, Bowen, Mayall and others – paid their deepest respects to the late scientist in many scientific periodicals. The Hubble archive displays a sheet of paper with verses written by a goose quill. These verses were written by Edward John Planckett, 18th Earl Dunsany, on the occasion of the death of the surveyor of the universe:

In memory of Dr Edwin Hubble

Now close the eyes that looked beyond the bound
Of human knowledge through the Milky Way
To clusters that we had not known to stray
Through that dark emptiness that rings us round;
And floating in those clusters he had found
Millions of suns like that which makes our day,
Where from his mountain window, ray on ray,
He saw the night with universes crowned.
Proud may we be that Man has seen so far.
And yet it never was his wont to boast
Of ought that he had seen beyond Earth's coast.
He would have known the wonder of a star,
And how those myriads opening to his eye
With all their lights taught Man humility.

The news of the death of the outstanding astronomer of this century rapidly spread among the family of astronomers the world over. At the meeting of the British Royal Astronomical Society, its president Dr Jackson said:

As Dr. Hubble delivered the George Darwin lecture here only a few months ago, the news of his sudden death has come to us as a shock. He will go down to history as one of the most distinguished astronomical observers of all time... It was his fortune as well as that of our science that the great telescopes required for these problems were available to him. His work on the nebulae will cause his name to be linked with that of Herschel who preceded him by more than a century.

On the other continent, independently but as if paraphrasing these words, Mayall said:

Perhaps the perspective of time is needed for a less subjective view, but it is tempting to think that Hubble may have been to the observable region of the universe what the Herschels were to the Milky Way system, and what Galileo was to the Solar system.

Years later, Mayall repeated his evaluation, this time with absolute certainty, in his biography of Edwin Hubble. However, Alan Sandage may have expressed his feelings in the most complete and precise manner when he referred to Hubble as the greatest astronomer since the time of Copernicus. Those words opened our book.

To quote Sandage:

No one knows where Hubble was buried. There is a certain mystery in what has taken place. There was no memorial service, there was no formal funeral. His wife was very protective of the whole situation... Milton Humason was possibly the only person who ever knew what has taken place after Hubble's death.

Nowhere on Earth are there monuments or memorial tablets in Hubble's honour, neither at Mount Wilson, nor at Mount Palomar.

Many years after his death, astronomers attached Hubble's name to an object in the skies. His name was given to one of the craters on the Moon, close to Mare Marginis. Asteroid No. 2069 Hubble moves between the orbits of Mars and Jupiter. The asteroid with the next number was given the name of Humason, with whom Hubble had worked for a long time on the problem of the red-shift.

Hubble's name was also given to an artificial object – the space telescope. In spite of unfortunate technical mishaps, astronomers expect great achievements from this instrument, whose mirror is almost as large as that of the 100-inch, but which is placed beyond the terrestrial atmosphere. As for the 100-inch reflector which made Hubble famous and was made famous by Hubble, its operation has been terminated. The Carnegie Institution was unable to finance its work any more.

On the night of 25 June 1985, a sad group of astronomers gathered in the telescope tower. They focused the telescope, perhaps for the last time, on Vega, the star at which Hale looked for the first time through the eyepiece of this instrument in distant 1917. The front page of the journal *Sky and Telescope* showed, as a symbol of the sad event, a photograph of the last observer, the British astronomer Roger Griffin with a board in his hands: 'The 100-inch telescope closed'. The era of this excellent instrument, with which the greatest discoveries of the first half of the 20th century had been made, was over. It 'outlived' Hubble by 32 years.

The true heroes of human history, culture and science do not die when their physical life ends. Having transferred their achievements, discoveries and ideas to the new generations, they gain immortality for centuries to come. Edwin Hubble was among such heroes of science of our century, which has indeed been rich in outstanding achievements. Hubble left us an impressive heritage – the discovery of the world of galaxies whose evolution obeys the law given Hubble's name; it would be logical to call it the Hubble Universe. With every new year, we understand more profoundly the importance of his accomplishment.

The second part of the book describes the fundamental results obtained after Hubble's time and new research projects.

PART TWO

Hubble's work continued

Distances to galaxies and the Hubble constant

Years and decades passed since Hubble's classic work discovered the expanding universe, and it became gradually clearer that this discovery had stimulated astronomers to tackle tremendous problems which were waiting for new generations of both observers and theorists.

Among the toughest problems that opened themselves to astronomers who continued Hubble's work was the task of determining the distances in extragalactic astronomy and the related problem of measuring the most important characteristic of the universe: the Hubble constant.

A sufficiently accurate evaluation of distances to galaxies, especially very distant ones, is extremely difficult since the distances are truly gigantic. It is not really surprising that even today we still know extragalactic distances only with considerable error, perhaps to within a factor of two. This uncertainty will immediately explain itself if the reader recalls that distances to stars even in our galaxy (with the exception of those nearest to our planet, distances to which are evaluated by the trigonometric parallax method) are only known to within several tens of percent.

Both in Hubble's time and today, the scale of extragalactic distances is established in several stages, as astronomers reach increasingly distant objects and build a sort of scale staircase. Each step of this staircase is characterised by methods specific to this step, and uses specific distance indicators.

Such indicators are objects with sufficiently well-determined luminosity (absolute stellar magnitude). The distance is then found photometrically, by measuring the apparent brightness of an object. In a different method, objects with known linear size can be chosen as such indicators. In this case, the measurement of apparent angular size in the sky also yields the distance.

Obviously, the difficulty lies in measuring the luminosity or linear size of indicators; in astrophysicists' lingo, the difficulty lies in calibrating them.

Typically, the sequence of measuring ever greater distances may be outlined as follows.

The first step is to measure the distance to the Hyades, one of the nearest loose clusters. This distance has been measured quite reliably, by a geometric method, and is found to be about 45 parsec. Knowing the distance to the Hyades and measuring the apparent stellar magnitude m, we can calculate the absolute stellar magnitude M of all member stars in the cluster.

The magnitude M of a star in whose central region hydrogen is converted into helium by nuclear fusion reactions is implied unambiguously by the colour of the star.

If we plot a diagram of the apparent stellar magnitude as a function of colour for the stars of a cluster (the Hertzsprung–Russell diagram), these stars form a chain known as the main sequence; they are readily distinguished from other stars. Observing now stars of the main sequence in other clusters, we can evaluate their respective M from their colour, then compare with their magnitudes m to find the distances to the clusters. This is how distances to clusters are measured in our galaxy. Some clusters contain Cepheids. Their absolute stellar magnitudes M can be found from m and the distance. The secret lies in Cepheids obeying the period versus absolute magnitude M relation. Now, with M obtained for at least several Cepheids, we can assume as known what period corresponds to what absolute stellar magnitude. Astronomers say that this procedure calibrates the relation. Cepheids are very bright stars ('supergiants') and can be seen as far away as the nearest galaxies. They are known as primary distance indicators. If a Cepheid is found in another galaxy, the distance to it and, consequently, the distance to the entire galaxy, is calculated by comparing the stellar magnitude m with M (determined by measuring the period of the Cepheid). Unfortunately, Cepheids are resolved only in the nearest galaxies (at distances up to several million parsecs). In order to move still farther, the next step must be taken: to find secondary distance indicators, that is, indicators of higher luminosity than that of Cepheids. The brightest stars of galaxies or the brightest globular clusters are among objects used as secondary indicators. Observations show that the absolute magnitudes of each of these types of indicators are acceptably identical in galaxies that belong to the same galaxy type.

The magnitudes M of the secondary indicators are found ('calibrated') by observing them in the nearest galaxies lying at already known distances. The secondary indicators make it possible to measure distances to the nearest clusters of galaxies (these are distances of the order of ten million parsec). Finally, third-order indicators are used to allow one to measure distances still further out into the universe.

As such indicators, one may use supernovae at their maximum brightness, or the brightest galaxies in clusters.

Other types of indicators are sometimes used. For instance, novae at their maximum brightness serve as primary indicators, linear diameters of ionised hydrogen clouds serve as secondary indicators, and so forth.

Obviously, one always has to make corrections for absorption of light in interstellar space and for a large number of other technical factors that we have no space to discuss here.

Just as obviously, errors inevitably accumulate on each step of this long staircase.

It is not surprising, therefore, that the first evaluations of distances contained considerable systematic errors, and that even now the uncertainty of the scale of extragalactic distances is still quite high.

Discrepancies pointing to inconsistencies in the evaluated distances to various objects became gradually more evident even when Hubble was still working.

For instance, variable stars of the RR Lyrae type were not found in the galaxies nearest to us, that is, the Large and Small Magellanic Clouds. This fact signified that they are so faint that photographs taken using available telescopes could not record their images. If these galaxies were at a distance of about 30 000 parsec (the estimate accepted at the time), RR Lyrae type stars would definitely be observable. Hence, they would have to be farther away than astronomers had believed.

Furthermore, current estimates of distances, based on Cepheids, indicate that the Andromeda nebula is at a distance of about 300 000 parsec from us. This galaxy is of the same type as our galaxy but the luminosities of globular star clusters and novae in it were found to be considerably lower than the luminosities of similar objects in our galaxy if this distance was used for evaluation. Also, the size of the Andromeda nebula did not correspond to that of our galaxy: it was significantly smaller even though both galaxies belong to the the same class. All this pointed to the Andromeda nebula being substantially more distant than astronomers had thought.

On the other hand, the French astronomer X. Mineur reconsidered the distances to Cepheids within our galaxy and concluded in 1944 that their absolute magnitudes are likely to be 1.5^m higher than had been assumed since Shapley's work.

However, all these indications were disregarded by astronomers until 1952 when W. Baade told the General Assembly of the International Astronomical Union in Rome that his measurements showed Cepheids to be 1.5^m brighter than was then generally believed and that the entire scale of extragalactic distances must be nearly doubled: indeed, the secondary indicators for greater distances had been calibrated on the basis of the nearest galaxies.

Baade's conclusion was confirmed in A. D. Thackeray's work performed at the same time, and then in numerous later publications.

In fact, the reconsideration of the extragalactic distance scale was to be resumed later.

Beginning in the middle 1950s, Hubble's former student, the American astronomer A. Sandage, and his colleagues worked relentlessly on re-measuring extragalactic distances. Sandage obtained more accurate values for distances to the nearest galaxies. They proved to be three times more distant than Hubble had thought. Furthermore, Sandage found a mistake in the work of his teacher. He was able to show that the brightest point-like images on photographs of sufficiently distant galaxies beyond the Local Group were not stars – as Hubble had classified them – but huge clouds of ionised hydrogen (now known as HII regions). Hubble was unable to distinguish between these objects and stars since, being so distant, they appear as dimensionless dots on plates. Only when plates of sufficiently high sensitivity in the red end of the spectrum became available, could Sandage identify the clouds by comparing photographs made through blue and red filters.

Compact HII regions in galaxies in the Virgo cluster were shown by Sandage to be brighter by 1.8^m than the brightest stars. Therefore, when Hubble evaluated the factor by which the Virgo cluster was more distant than the galaxies nearest to us and used the brightest stars for that, he actually worked with HII regions and underestimated the distance to the cluster by a factor of about two. It was also found later that the highest-luminosity stars in galaxies are about 25 times as bright as had been thought in Hubble's time. In the first half of the 1970s it was accepted that if corrections to the scale of distances to the nearest and then more distant galaxies were introduced, the Virgo cluster and still more distant ones would be shifted six to ten times farther away than Hubble had placed them. Correspondingly, the evaluation of the Hubble constant H had to be reduced by the same factor. From 500 km/s per megaparsec as given by Hubble, it dropped to 50–100 km/s per megaparsec.

In 1968, Sandage used the brightest galaxies in large clusters as distance indicators. This approach allowed him to penetrate to distances at which the velocity of recession reached 140 000 km/s ! The Hubble constant determined in this way decreased to $H = 75$ km/(s.Mpc).

During the last ten to fifteen years, work on the scale of extragalactic distances and the Hubble constant has been carried on at least as intensely as in preceding decades. New methods have been suggested and older ones constantly elaborated and corrected.

Among new techniques, we need to point out the relation, discovered by R. Tully and R. Fisher in the USA, between the width of the spectral radio line of neutral hydrogen (at 21 cm wavelength) and the luminosity of a galaxy. The linewidth is determined by the velocities of motion of the gas, and these velocities depend, in their turn, on gravitational forces and, hence, on the mass of the galaxy. This is the explanation of the discovered relation, which permits a calibration of the absolute magnitudes of galaxies.

Using this relation, one can determine the luminosity of a galaxy from the measured width of the hydrogen radio line, and then calculate the distance by comparing the luminosity with the apparent magnitude.

The Tully–Fisher method does not apply to elliptical galaxies since they do not contain large amounts of neutral hydrogen. A similar method of calibration of their luminosities can be used, if one calculates from the spectral data the velocities of motion of stars, rather than of the gas.

Another promising method is to determine jointly the linear velocity of expansion of the envelopes of the exploding supernovae in the galaxies (this is done by measuring the Doppler shift of the lines in their optical spectra) and to measure the rate of growth of the angular size of the receding envelopes. The latter measurements are carried out with the most modern very-long-baseline radio-interferometers which provide angular resolution better than 10^{-4} seconds of arc. A comparison of the linear with the angular velocity gives the distance to the supernova and, hence, to the parent galaxy.

Another method was proposed in the USSR by Ya. B. Zeldovich and R. A. Sunyaev. It is based on simultaneously observing X-ray radiation of the hot gas in clusters of galaxies and the relic microwave background radiation of the hot universe, scattered by this gas (the relic radiation and its discovery are discussed on pp. 143–51).

Even though new methods led to a number of detail corrections, they did not result in any substantial increase of the reliability of distance measurements.

Another problem arose in evaluating the Hubble constant, in addition to difficulties of measuring large distances. Beginning in the 1960s, it was becoming increasingly clear that gravitational forces due to large local conglomerations of matter (clusters of galaxies) appreciably affect the motion of relatively near galaxies receding at velocities of not more than 4000 km/s. These forces can substantially distort, over a relatively small local scale, the overall Hubble expansion. Such local distortions must be specially taken into account when evaluating the Hubble constant.

In the 1980s and especially very recently, at the beginning of the 1990s, it became obvious that considerable deviations from the 'uniform Hubble flow' also exist. The most conspicuous example is 'The Great Attractor'. A group of seven astronomers (known as 'The Seven Samurai') discovered in the second half of the 1980s that our galaxy participates, together with its surrounding galactic populations, in a common motion towards the Hydra and Centaurus constellations. This additional motion is superposed on the uniform Hubble flow. It involves a region of more than 50 Mpc in diameter and is characterised by velocities of about 600 km/s. It is assumed that these peculiar velocities are caused by the attractive pull of a region of high concentration of matter, of total mass up to 10^{16} solar masses (hence the name Great Attractor). Features of even larger scale in the observable universe were found later.

Leading specialists on extragalactic distances have split during recent decades into two clearly defined groups. One of them, headed by Alan Sandage and the Swiss astronomer G. Tammann, insists that the Hubble constant is approximately $H = 50$ km/(s.Mpc).

Sandage and Tammann wrote in 1986:

... we adopt $H_0 = 50 \pm 7$ [*the numbers are given in units of km/(s.Mpc); sometimes astronomers attach the subscript '0' to H, emphasising that the contemporary epoch in the evolution of the universe is meant*] as the global value of the Hubble constant, based on the present assessment of the evidence to date. We take this to be an *upper* limit because all systematic errors of the nature of statistical bias always work in the direction that the true distances are larger than the inferred distances calculated when the biases are neglected. However, we also caution that much is unknown yet on basic questions of internal absorption in the relevant galaxies, and on *systematic* errors in distances to the *local calibrators*. The problems of zero-point of the Cepheids remain, and cry for a modern solution.

However, quite a few astronomers reject this conclusion. Advocates of the second group of experts on extragalactic distances are of the opinion that the Hubble constant lies somewhere near $H = 100$ km/(s.Mpc). The leader of this group, the French astronomer G. de Vaucouleurs, concluded in his review paper of 1982 that the most probable value of the Hubble constant is $H = 95 \pm 10$ km/(s.Mpc), or, in a more general form, $116 \geq H \geq 81$ km/(s.Mpc)). The former group of astronomers is often referred to as advocates of the 'long' scale of extragalactic distances, while the latter group insist on the 'short' scale.

What is the reason for such large discrepancies in the estimates of H by the best experts? There is no doubt that the contradiction is based on the insufficiency of our knowledge, caused by the enormous difficulties of the problem. More specifically, the difference in conclusions is a consequence of differences in the methods used by the two camps of scientists.

The main difference between their methods is that Sandage and Tammann choose the least possible number of – presumably – most reliable distance indicators (primary, secondary, etc.) and an equally minimal number of reliable techniques of calibrating them ('Principle of Precision Indicators'), while G. de Vaucouleurs, S. van den Bergh and other astronomers prefer to take a larger number of indicators and to calibrate them by all possible methods. In terms of a colourful expression used by de Vaucouleurs, the former group prefers to put all their money on a single horse, while the latter group preaches the philosophy of 'spreading the risk'. It is necessary to emphasise that the difference between the 'short' and 'long' distance scales reaches a factor of two only for distances to the most distant objects, which are much greater than distances to the nearest large galaxy cluster in the Virgo constellation. Both groups give almost identical, or at least not very different, estimates of distances to the nearest galaxies in which Cepheids are found – these are the most reliable primary indicators of extragalactic distances. Here the differences are only about 20%. As the distances increase, the discrepancies between the scales grow and reach a factor of 1.5 at the distance of the Virgo cluster and approximately 2 for a much more distant cluster in the constellation Coma Berenices.

When speaking about the two estimates, $H = 50$ and $H = 100$ km/s per megaparsec, one has to remember that they imply different estimates of the age of the universe. For instance, $H = 50$ km/(s.Mpc) corresponds in the simplest Einstein–de Sitter model to the time $t = 13 \times 10^9$ years (13 billion years), while $H = 100$ km/(s.Mpc) corresponds to half that duration. In contrast to this, estimates of the age of globular clusters typically exceed 15 billion years, reaching 18 billion years. Uncertainties in the age of these objects, which are probably the oldest in the universe, are not smaller than those of the Hubble constant. Nevertheless, it appears to be very difficult, or even impossible, to reconcile an age for globular clusters of about 15 billion years, even remembering the possible error, with the age of the universe of 6.5 billion years (in the simplest cosmological model) which is implied by the value of the larger Hubble constant, $H = 100$ km/(s.Mpc).

In all likelihood, cosmological theory and observational results can be reconciled only be reviving Einstein's original idea on repulsive forces described by the cosmological constant Λ in the field equations. In principle, introduction of these forces allows us to 'stretch' the time since the beginning of the expansion of the universe to any desirable length of time.

Summarising the results of the work on elaborating the value of the Hubble constant, the American astronomer P. W. Hodge wrote:

There is no general consensus among these authors, except that virtually all find values of H_0 that lie in the range 50 to 120 km.s^{-1} Mpc^{-1}. The large degree of disagreement is clearly the result of the uncertain nature of various assumptions involved in each step of the process. Many of the fundamental data are both less accurate and less well understood than originally assumed and much of the hoped-for uniformity in nature being found to be less than perfect as far as 'standard candles' are concerned.

Even though the numerical value of the Hubble constant is not known with satisfactory accuracy, the fact of the proportionality of recession velocity to distance has been established very reliably. It is not mandatory to know the value of the coefficient H in order to test the law describing the expansion of the universe. Indeed, if we know that a distance indicator has a constant luminosity (even if it is known with substantial error), a comparison of the apparent magnitudes of these objects located at different distances yields directly the ratio of their distances. The apparent magnitudes of such indicators thus give a measure of relative distances.

Hubble was the first to propose that whole galaxies found in clusters be used as distance indicators. The lone galaxies that are occasionally found cannot be used for this purpose since their luminosities are very different. The total luminosity of stars in some galaxies is tens of times greater than the luminosity of our galaxy. However, some are hundreds of times fainter. As for the brightest galaxies in clusters, they typically have fairly similar luminosities (in fact, Hubble himself used not the brightest galaxy in a cluster but the fifth-brightest, but this does not change the situation at all). Therefore, the proportionality of the expansion rate of the universe to distance is tested by plotting the apparent magnitude of the brightest galaxies in clusters as a function of the red-shift which characterises the recession velocity of the cluster. This last quantity is usually denoted by the letter z; it is defined as the ratio of the change in the wavelength of spectral lines to the wavelength of the same line emitted by a stationary source: $z = \Delta\lambda/\lambda$.

Figure 3 (p. 74) shows this relation using the data of A. Sandage and G. Tammann of 1981. We see that the logarithm of z and the observed magnitudes fit quite accurately the linear dependence, predicted by Hubble's Law, up to $z \approx 0.7$.

The left-hand corner of the figure shows a black rectangle. It corresponds to the region of data available to Hubble in 1929 when he had formulated his law. This comparison clearly demonstrates the impressive progress achieved by astronomy in the last fifty-odd years.

It seemed quite recently that progress could be even greater. This expectation was connected with the quasars discovered at the beginning of the 1960s. The history of the discovery of these objects began in September 1960 when T. Matthews and A. Sandage, working with the famous 200-inch telescope, obtained a photograph of a very compact radio source whose number in the Third Cambridge Catalogue was 3C 48. The object looked like a 16^m star enveloped in a faint nebula. A month later, Sandage recorded its spectrum. It showed broad emission lines which could not be identified with lines of any known chemical element. In the following two years, Matthews and Sandage and then other astronomers were able to show that some other point-like radio sources could also be identified with faint star-like objects in optical radiation.

In 1963, the Dutch astrophysicist Maarten Schmidt, working at the Mount Palomar observatory, obtained the spectrum of a star-like optical image of the compact radio source 3C 273. The object appeared on the photograph as a 13^m star. Schmidt established that the unusual emission lines in the spectrum of the source belong to hydrogen, the most abundant element in the universe, but are shifted to the red end of the spectrum, to $z = 0.16$. Large red-shift can only be caused by high-speed recession of the object due to the expansion of the universe. It is not difficult to show, using Hubble's Law, that the object is extremely distant and that its luminosity is hundreds of times more than that of the largest among galaxies. Large red-shifts were soon found in other similar objects. All of them were later given the name *quasars*.

The discovery of these fantastically powerful sources of radiation in the universe was already a sensation. However, the extremely small size of these incomparably powerful sources was even more staggering. The size of quasars was evaluated when their brightness was found to vary.

The news of the unusual objects reached Moscow in the first days of March 1963. On I. S. Shklovsky's initiative, Yu. N. Efremov and one of the authors of this book (A. Sharov) studied the brightness of the 3C 273 quasar using the plates of the P. K. Sternberg Astronomical Institute in Moscow. It was established that its brightness varied with an amplitude of 0.7^m. At the same time, the American astronomers H. Smith and D. Hofflighter also measured the variability of 3C 273 and reported its amplitude as 0.6^m. Incidentally, the papers of the Soviet and American astronomers were sent for publication on the same day, 9 April 1963.

It was found that sometimes the brightness of 3C 273 changed appreciably over several days. This meant that the linear size of the quasar could not be more than several light days. If light rays from different parts of the object reach us simultaneously, they have been emitted at very different moments: earlier from more distant parts and later from less distant parts. If quasars were larger, variations would average out and it would be impossible to observe variations longer then several days in the observed total incident flux.

The nature of quasars puzzled astronomers for a long time. Gradually, it became clear that quasars are nuclei of very distant giant galaxies.

Nuclei are found in the central part of quite a few galaxies but their luminosity is typically not very high. For some reason, nuclei of quasars are in a state of extreme excitation; their luminosity reaches 10^{45}–10^{47} erg/s, which exceeds the luminosity of the largest galaxies by one or even two orders of magnitude.

It would be natural and very appealing to make use of such powerful emitters, seen at such tremendously large distances, to test Hubble's Law and to solve other cosmological problems. It became clear, however, that this is extremely difficult. First of all, luminosities of quasars are scattered in an immensely wide range, in contrast to the brightest galaxies, and thus cannot serve as distance indicators. Furthermore, the luminosity of quasars varies greatly with time. The characteristic lifetime of a quasar is likely to be of the order of 10^7 years. The age of galaxies is of the order of 10^{10} years. The much shorter period of activity of quasars in comparison with the age of galaxies results in relatively rapid variability of quasar luminosity.

The evolution, that is, the variability of luminosity and of other properties of galaxies and quasars, becomes one of the most important problems when we turn to objects red-shifted to $z \approx 1$ or greater. Light arriving with us today has left these distant objects billions of years ago when they were substantially younger and thus the luminosity of our standard indicators had been very different from those we observe in nearby objects. In addition, we understand only poorly the evolution of galaxies and quasars.

In the meantime, it is extremely important to study the 'apparent magnitude versus red-shift' dependence for objects with red-shift z greater than unity, since a number of important factors characterising our universe affect the curve at such large distances. First, it is important to remember that we see the universe in its very distant past when the Hubble constant was different: indeed, the expansion is slowed down by the gravitational forces of matter. The observation of objects with large z would make it possible to determine the rate of deceleration and thus calculate the mean density of matter in the universe. Second, relativistic effects, such as changes in the flow of time in strong gravitational fields and the curvature of space in the universe, are already appreciable for such distances. Hubble realised, and stated, that in principle, such observations allow one to measure the anticipated effects. In fact, these effects are 'mixed' with the effects of evolution; separating them is an extremely difficult task. Unfortunately, one has to recognise that the observation of very distant objects has not led so far to any definite conclusions; astrophysicists pin all hopes on future studies which we will outline in the next section.

The largest red-shifts of galaxies and quasars measured by the end of the 1980s were $z_g = 3.8$ and $z_q \approx 5$, respectively. Readers will recall that the largest red-shifts available to Hubble did not exceed a mere $z = 0.004$.

Current and future research projects

Projects that involve launching astronomical instruments into extraterrestrial space occupy a very important place among the many projects designed to enlarge substantially our knowledge of the large-scale structure of the universe and of its evolution. We will describe here some of the projects which directly continue the line of research initiated by Hubble.

According to plans, in 1986 NASA was to put in orbit around the Earth a space telescope with a mirror 2.4 m in diameter: the Hubble Space Telescope (HST). The tragic catastrophe of the Challenger space shuttle delayed the project but did not cause its cancellation. The satellite with the Hubble telescope and all the observation instruments was finally launched in April 1990. The total cost was 1.6 billion dollars. This unique instrument was designed to bring optical astronomy to a qualitatively new stage; the event was to be at least as significant as was the beginning of operations of the 200-inch instrument. A special Space Telescope Science Institute was created in Baltimore to organise the work of the telescope in orbit.

The mirror of the new instrument was designed to obtain images of unprecedentedly high quality of celestial objects, not perturbed by the terrestrial atmosphere. The telescope was to observe objects over a very wide range of wavelengths of electromagnetic radiation, from far ultraviolet to far infrared.

Its special instruments of extreme importance for extragalactic observations are: (1) the Wide Field Camera (WFC) with a field of view of $2.7' \times 2.7'$, composed of 1600×1600 photon detectors (charge coupled devices) for studying objects of magnitudes from 9.5 to 28; (2) the Faint Object Camera (FOB) for very faint objects, with $11'' \times 11''$ field of view and angular resolution of $0.02''$ for the observation of objects in the range of magnitudes from 21 to 28; and (3) a spectrograph for very faint objects, down to a magnitude of almost 26. These instruments were to perform detailed photometric monitoring and spectral investigation at high angular resolution of objects located almost ten times farther away than could be detected if observed through the atmosphere from the surface of the Earth.

The possibility of studying objects much fainter than those accessible with the largest Earth-bound instruments is the decisive factor. The astronomers of the 'long-scale' and 'short-scale' groups arrive at almost the same distances to galaxies in which Cepheids are now resolvable. The differences accumulate when the two groups turn to more distant galaxies, in which Cepheids are out of reach and one has to resort to working with considerably less reliable secondary, and then tertiary, distance indicators. This shows the importance of measuring the distance to the nearest galaxy cluster in the Virgo constellation via Cepheids, without using secondary indicators.

The Hubble Space Telescope was expected to solve this important problem of observational cosmology. Cepheids with periods of about 20 days and average absolute stellar magnitude of -5.5^m in the galaxies of the Virgo cluster must have an apparent magnitude of 26^m and thus should be detectable after exposures of about 50 minutes.

The observation of numerous galaxies in this cluster was to allow the Space Telescope scientists to find and calibrate secondary distance indicators as well, to a much better accuracy than had been possible before. Such secondary indicators could be resolvable from distances almost ten times greater than those we work with now. One or two exposures are sufficient to find and study these indicators, HII regions and the brightest globular clusters, at distances up to that to the globular cluster in the constellation Coma Berenices, which is about six times farther away than the Virgo cluster.

It was expected that all this progress would substantially improve our knowledge of the scale of the universe and make it possible to measure the Hubble constant with an accuracy of about 10%.

The Space Telescope could also greatly improve the calibration of the primary indicators in our galaxy. Namely, it was planned to apply the direct trigonometric parallax method and measure to within 10% distances to stars 100 parsec from us. This would be an obvious improvement for distances to the nearest stellar clusters, those which form the basis of the long staircase of the space scale system.

Sufficiently accurate data on distances to objects tens of millions of parsecs removed will allow astronomers to measure reliably small deviations in the motion of galaxies: those peculiar motions imposed on the general expansion of the universe and caused by the gravitational pull of large clusters. The measured deviations permit calculation of gravitational forces and of the aggregate mass of matter in clusters of galaxies, including forms of matter which are difficult to observe. Readers should recall that poor knowledge of this quantity is the weakest link in calculating the average density of matter on a very large scale. Some day, progress on this question will lead to solution of one of the most difficult problems in observational cosmology: the determination of the average density of matter in the universe. This quantity will inform us whether the universe will expand indefinitely or whether the expansion will eventually be replaced by contraction.

Obtaining the dependence of the apparent stellar magnitude of the brightest galaxies on the value of the red-shift for clusters with $z > 1$ will serve the same purpose. The deviation of this curve from a straight line gives information on the average matter density in the universe and on the curvature of three-dimensional space.

All these observations were planned for studying very distant objects that we observe in their past; the light we receive left them billions of years ago. These observations were also expected to reveal the evolution of galaxies and quasars over such long stretches of time. Such data are extremely important for understanding the history of the universe.

Such were the plans. We described them in the Russian edition of this book which appeared in 1989. So much hope was connected with that launch ...

Enormous difficulties are always encountered when huge projects are implemented, and there is a danger of very serious failures. The Hubble Space Telescope had an extraordinarily large share of bad luck. The American astronomer J. Huchra, when discussing plans for improving our knowledge of H several years before the planned launch, exclaimed vehemently: 'Keep your fingers crossed for the Hubble Telescope'.

The satellite with the telescope was finally launched in April 1990, and in June of the same year the National Aeronautics and Space Administration acknowledged that the image at the focus of the main mirror was blurred. It was found that owing to unfortunate errors and negligence during manufacturing and testing of the mirror, it had spherical aberration ten times greater than admissible. Later some gyroscopes broke down on the satellite in orbit, and one of the scientific instruments – the Goddard High Resolution Spectroscope – was handicapped. The news of the spherical aberration, which blurs images, was a shock to astronomers. One can never forget the huge sums spent on constructing the telescope. The aberration of the main mirror meant that some of the planned observations could not be run at all, while others would consume considerably more time for implementation.

Plans for 'repairing' the telescope started to appear. In the meantime, some successful observations were nevertheless carried out.

As far as determination of the value of H is concerned, observations of the ring of hot gas around the remnant of the 1987 supernova in the Large Magellanic Cloud proved to be of great interest. This gas was ejected from the progenitor star several thousand years before the supernova flared up; during the explosion, the ring was heated up by radiation and started to glow. By comparing the angular size and visible 'flatness' of this ring as measured by the Hubble Telescope with observations obtained with the International Ultraviolet Explorer Satellite, it was possible to calculate the distance to the supernova, and hence, to the Large Magellanic Cloud. This distance was found to be 169 000 light years, determined with an accuracy of 5%. As readers will remember, greater precision in measuring the distance to the nearest galaxies is an important element in constructing the scale of extragalactic distances.

A large number of studies carried out using the Hubble Telescope, especially spectrographic studies, began to appear regularly in journals in the second half of 1991.

As for plans for repairing the Hubble Telescope, the most promising is the project of installing a system of correction lenses to cancel out the spherical aberration. These correcting lenses can be mounted on the telescope by astronauts during the space visit planned in the next few years.

Astronomers still hope that the Hubble Telescope will be able to do its job despite the cascade of disasters.

Completing the story about the potential of the Hubble Space Telescope in establishing the extragalactic distance scale, we want to emphasise the following aspect. When using the Hubble telescope to solve the problem of determining the structure of the universe, astronomers continue work that Hubble initiated in the 1920s. When the distance scale is established with an accuracy of 10 percent, this will have completed work on one of the greatest problems formulated as a result of Hubble's efforts. It is very symbolic that the Hubble telescope continues the work started by Edwin Hubble himself.

Whatever the promise of the Space Telescope in improving the scale of extragalactic distances, we cannot be completely satisfied with the accuracy anticipated by its designers. Quite a few problems in cosmology will not be solved. For example, we will hardly be able to clarify whether our universe is open (permanently expanding) or closed (gradual slowing down of expansion and ultimate onset of contraction). The problem is that according to theoretical estimates, the difference between the mean matter density of the universe and its critical value is likely to be many orders of magnitude less than the maximal accuracy with which this parameter can be determined by observations with the Hubble Space Telescope.

The method of photometric distance indicators proposed and actively applied by the pioneers of the investigation of the universe will hardly be able, in view of the underlying principle, to improve accuracy above the value expected of the Hubble telescope.

A new step in performing measurements on the immense spaces of the universe will become possible only when extragalactic astronomy is able to apply the direct technique of distance measurement: the method of trigonometric parallax.

So far, only distances to the nearest stars can be measured in this way.

Can the sensitivity of the method be increased million-fold, as required by extragalactic astronomy? N. S. Kardashov, Yu. N. Pariisky and N. D. Umbaraeva in the USSR showed in the 1970s that measuring distances to galaxies and even to the very boundaries of the observable universe by trigonometric techniques is in principle possible. The method will become feasible when space radio-interferometers with bases of the order of the diameter of the Earth's orbit are launched.

The angular resolution of a radio-interferometric system, and the possibility of measuring parallaxes determined by it, is a function of the ratio of the wavelength of the electromagnetic radiation used by the system to the base length, that is, the distance separating the radio-telescopes employed to synthesise the interferometric pattern. If cosmic radio-telescopes worked at a wavelength of one centimetre and the distance between them were about 300 million kilometres (the diameter of the Earth's orbit), the angular resolution would reach 10^{-10} seconds of arc. This would be sufficient for measuring distances up to several billion parsecs, that is, up to the boundaries of the universe observed now by optical instruments!

Of course, there is still a very long way to go before implementing such projects and many difficulties of a technical and theoretical nature will have to be overcome. First, at least three radio-telescopes would have to work simultaneously (one of them can be on the Earth), each having instruments for extremely precise measurement of the distances and relative velocities of the telescopes. Furthermore, it is necessary to find suitable objects in distant galaxies: very compact objects, less than the diameter of the Earth's orbit, emitting with sufficiently high power in the radio-frequency band. Finally, it would be necessary to take into account numerous sources of possible deviations in the propagation of radio waves in the course of their long journey to the observer.

The first steps in designing space radio-telescopes have already been made. In 1976, the space radio-telescope KRT-10 functioned on the Soviet space station Salyut 6. In 1986, American scientists created an interferometric system using a radio-telescope on a satellite. Very important studies are planned for the future.

The Astro Space Centre of the P.M. Lebedev Physical Institute in

Moscow plans to implement the project RADIOASTRON in the near future. This is a programme of constructing gradually more complex and multi-purpose space radio-telescopes during the next twenty years.

According to this programme, a space radio-telescope with an antenna 10 m in diameter is to be placed in orbit within this decade. The telescope will work on frequencies from 0.3 to 22 GHz. Together with Earth-based radio-telescopes, it will form an interferometric system with base length up to 80 000 kilometres. In the next five years, a similar instrument will be launched, working in the millimetre wavelength band – at frequencies from 22 to 230 GHz. Finally, a space radio-telescope with an antenna 30 m in diameter, working at frequencies from 1.7 to 230 GHz, is to be designed and launched in the next five years.

The realisation of this programme will be a significant step to implementing a dream of cosmologists: to achieve triangulation of the entire observable part of the universe, like the triangulation of the globe completed in the past. As the latter triangulation led to measuring distances between remote points, to constructing exact maps, to finding the curvature of the surface of the globe and to calculating the size of the planet, so will the future triangulation of space produce a three-dimensional map of the surrounding universe and allow us to measure the curvature of space. In its turn, it will disclose new details about the history of the universe and add greater plausibility to predictions about its future.

As the creation of RADIOASTRON progresses, a number of other problems of observational cosmology will also be solved.

Even the first stages of the realisation of this project will make it possible to measure accurately the angular diameters of the receding envelopes of supernovae and to determine distances to the galaxies where these explosions take place, with an accuracy of up to 10% (for the nearest galaxies).

In 1986, N. S. Kardashov remarked that RADIOASTRON will provide data for finding the peculiar motions of galaxies, that is, their angular displacement on the celestial sphere due to their peculiar three-dimensional velocities of hundreds, and maybe thousands, of kilometres per second. Galaxies have these peculiar velocities in addition to the velocity of their regular recession due to the expansion of the universe. These observations will not only make a reconstruction of the large-scale motion of matter in the universe possible for the first time but will also provide a way to determine the average statistical parallaxes of whole ensembles of galaxies, similarly to what is done in stellar astronomy to evaluate distances to sufficiently distant stars in our galaxy. Another possibility arising from conducting high-accuracy angular measurements of very distant objects is the creation of long-baseline systems of optical interferometers.

This brief survey shows that much work needs to be done and that this work is already well-planned, which is just as important.

The discovery of the hot universe

Friedmann's theoretical prediction of a non-stationary universe and Hubble's discovery of the expansion of the universe were the first steps on a long and difficult path that leads to an understanding of how the universe 'exploded', the meaning of this unusual explosion that occurred about 15 billion years ago, and how the universe is organised now.

The next outstanding step on this path was the discovery of the hot universe.

Various physical processes were important at different stages in the expansion of the universe. The density of matter was tremendously high at the onset of cosmological expansion. The dominant processes at that time were very different from those we observe today. They predetermined the state of the world we see around us and, among other things, the existence of life. Can we say anything about the processes that dominated the first moments of expansion? Yes, we definitely can. Events during the first minutes after the 'Big Bang' had consequences that left such indelible 'fingerprints' that we can now reconstruct many of their features.

The most important of them were the nuclear reactions occurring during the very first moments at very high matter and radiation densities. These reactions later produced light chemical elements. Calculation of the characteristics of nuclear reactions allows us to predict the chemical composition of the matter from which celestial bodies were later formed.

There are two fundamentally different scenarios for the conditions under which matter began to expand in the universe: the matter could be cold, or it could be hot. The corollaries of nuclear reactions differ greatly in these two scenarios. Historically, the cold birth of the universe was considered first (in the 1930s).

At this time nuclear physics was very young. The first assumptions postulated that all matter in the universe initially existed as cold neutrons. It was later shown that this assumption led to contradiction with observational data. The difficulty lies in the following. After creation, a neutron in the free state decays, on average, in 15 minutes, transforming into a proton, an electron and an antineutrino. Protons created in the universe would bind to the surviving neutrons and form the nuclei of deuterium atoms. Later on, the chain of nuclear reactions would produce helium atoms. Calculations show that in this scenario, the formation of more complex nuclei is practically impossible. Hence, all matter in the universe would be converted into helium. This conclusion is in obvious contradiction with observations. We definitely know that most matter in the universe consists of hydrogen, not helium.

The observed abundance of chemical elements in the world rejects the hypothesis that matter first existed as cold neutrons.

In 1948, G. Gamow, and later Gamow with his colleagues R. Alpher and R. Hermann in the USA, published papers which suggested a 'hot' scenario for the initial stages of expansion of the universe. The main purpose of the authors of the hot universe hypothesis was to calculate the currently observed abundances of various chemical elements and their isotopes from the analysis of nuclear reactions in the hot matter at the start of cosmological expansion.

The tendency to explain the origin of *all* chemical elements by their synthesis at the very beginning of expansion was very logical for the physics of the 1940s. The problem was that the time elapsed since the Big Bang was erroneously evaluated at that stage as 2 to 4 billion years. This error followed from the overestimated value of the Hubble constant. Comparing the age of the universe of 2–4 billion years with the estimated age of the Earth of 4–6 billion years, one had to conclude that the Earth, the Sun and stars grew from the primordial matter with fully evolved chemical composition. It was assumed that this composition did not change significantly after the Big Bang since the synthesis of elements in stars is a slow process and there was simply not enough time for its completion before the formation of the Earth, the Sun and other bodies.

A subsequent re-evaluation of the scale of extragalactic distances entailed a reconsideration of the age of the universe. The theory of stellar evolution successfully explained the origin of heavy elements (those heavier than helium) by nucleosynthesis in stars. The desire to explain the origin of *all* the elements, including the heavy ones, at the initial stage of expansion could be discarded. The gist of the hot universe hypothesis was, however, correct.

On the other hand, analysis had demonstrated that the helium abundance in stars and interstellar gas was about 30%. This is much higher than can be explained by nuclear reactions in stars. Therefore, helium must have been synthesised, in contrast to heavier elements, at the beginning of the expansion of the universe. The main component of the universe is hydrogen. Its fraction by mass is about 70%. All other elements amount to only a small percentage.

The main idea of Gamow's theory was that the high temperature of matter prevents it from turning into helium entirely. About 0.1 s after the expansion had started, the temperature was about 30 billion kelvin. The hot matter contained numerous high-energy photons. The density and energy of photons were so high that light interacted with light, creating electron–positron pairs. The annihilation of pairs can create photons, and also neutrino–antineutrino pairs. Ordinary matter is immersed in this 'cauldron'. Complex atomic nuclei could not exist at very high temperatures. They would immediately be broken by surrounding high-energy particles. The heavy particles of matter are therefore neutrons and protons. Interaction with high-energy particles in the 'cauldron' makes neutrons and protons rapidly convert into one another. Reactions synthesising heavier particles from neutrons and protons do not take place because a deuterium nucleus is immediately destroyed by high-energy particles. Owing to the high temperature, the chain leading to helium formation is thus broken at the very first link.

Some amount of deuterium survives and initiates the synthesis of helium only when the expanding universe cools down to a temperature below one billion kelvin. Calculations show that at this moment the fraction of neutrons in matter is about 15%. Combining with an equal amount of protons, these neutrons form helium amounting to about 30% of matter. The remaining heavy particles become the nuclei of hydrogen atoms. Nuclear reactions stop after the first five minutes of expansion.

This is how the theory predicts formation of 30% helium and 70% hydrogen as the main constituents of primeval matter.

Analysis of various scenarios at the onset of cosmological expansion has not ended with Gamow's hypothesis. A very elegant attempt to revive the cold-birth scenario was made at the beginning of the 1960s by Ya. B. Zeldovich. He assumed that the cold matter initially consisted of protons, electrons and neutrinos. Zeldovich showed that expansion would transform this mixture into a purely hydrogen plasma. According to this hypothesis, helium and other chemical elements were synthesised later, after stars had already formed. Note that data on helium abundance in pre-stellar matter were still very uncertain in the 1960s.

The true picture would be very hard to deduce if theories of the early

Universe could be tested only by data on the abundances of the chemical elements. Indeed, it is not easy to decide how much helium has been synthesised in stars and how much in the early universe. Discussion could continue for a long time.

There is, however, another method of testing the theory. Gamow's theory predicted the existence of primordial (relic) electromagnetic radiation in today's universe.* This background is a remnant of the epoch when matter was dense and hot. This radiation has cooled down in the course of expansion; today its temperature was expected to be from 1 to 30 K. Electromagnetic radiation at such a low temperature is in the centimetre and millimetre wave bands.

It would be logical if astrophysicists had turned with interest to the prediction of the microwave background radiation made in the first papers of G. Gamow, R. Alpher and R. Hermann, and if astrophysicists had then attracted the attention of radio-astronomers who could try to detect this background.

Nothing of the sort happened, however. Science historians are still puzzled as to why no one attempted to search for the primordial radiation of the hot universe. Before turning to their guesses, we will retrace the sequence of events leading up to the discovery itself.

In 1960, an antenna was constructed in the USA to receive reflected radio signals from the ECHO satellite. In 1963, this antenna was no longer in use, and two radio engineers, Robert Wilson and Arno Penzias of Bell Laboratories, decided to use it for radio-astronomical observations. The antenna was a 20-foot horn reflector. At that time, this antenna equipped with new receiver electronics constituted the highest-sensitivity instrument for detecting radio waves coming to us from large areas of the sky. The radio-telescope was designed for measuring the radio emission of the interstellar medium in our Galaxy. Observations were conducted at the 7.35 cm wavelength. Penzias and Wilson were not hunting for the microwave background; they knew nothing about the hot universe theory.

All possible sources of error have to be taken into account if one wants to measure galactic radiation in the radio bands. Noise is caused by the generation of radio waves in the atmosphere of the Earth, it is emitted by its surface, by the antenna itself, and within electric circuits and receivers.

---

* This term was suggested by the Soviet astrophysicist I. S. Shklovsky. Another term is the cosmic, or background, microwave radiation.

All sources of noise were carefully analysed and taken into account. Nevertheless, Penzias and Wilson were surprised to find that whatever the direction in which the antenna was pointed, it reported some radio emission of constant intensity. This could not be radiation from within our Galaxy, because in this case the intensity would vary depending on whether the antenna looked in the plane of the Milky Way or perpendicularly to it. Furthermore, galaxies similar to ours would also emit on the 7.35 cm wavelength. No such radiation was detected. Two possibilities remained: either there was an unknown source of noise, or the radiation came from somewhere in cosmic space. There was a suspicion that the noise was coming from the antenna. A thorough check showed that the antenna was not to blame. Hence, the radiation came from space, from all directions and with equal intensity.

Further events that led to the solution of the puzzle were often accidental. When talking to his colleague B. Burke about quite different subjects, Penzias mentioned, by coincidence, the mysterious radiation detected by their antenna. Burke recalled that he had heard of a talk by P. Peebles, who worked with the well-known physicist Robert H. Dicke. In this talk Peebles described something like a residual radiation of the early hot universe which must now have a temperature of about 10 K. Penzias called Dicke and the two groups of scientists met. Dicke and his colleagues P. Peebles, P. Roll and D. Wilkinson realised that Penzias and Wilson had discovered the primordial microwave radiation of the hot universe. Dicke's group in Princeton was at that time discussing the development of equipment for similar measurements in the 3 cm wave band, but observations had not yet begun. Penzias and Wilson had already made their discovery.

In summer 1965, the *Astrophysical Journal* published the papers of Penzias and Wilson about the discovery of the microwave background and the paper of Dicke and his colleagues about its interpretation in terms of the hot universe theory. The first observations showed that the temperature of the microwave background was about 3 K.

In subsequent years, numerous measurements were carried out on various wavelengths from tens of centimetres to a fraction of a millimetre.

Observations proved that the spectrum of the microwave background radiation fitted Planck's formula, as could be expected for radiation with a definite temperature. Spectral data confirmed that the temperature of the radiation was quite close to 3 K.

This is how an accident led to the great discovery of this century, proving that the universe was indeed hot at the beginning of the expansion. For this discovery, which was a significant continuation of Hubble's work, Arno Penzias and Robert Wilson won the Nobel Prize for physics in 1978.

Today, the observation of the relic background, of its spectrum and degree of isotropy, provides very important information on the large-scale structure of the universe, especially about its past history. However, a discussion of this aspect would lead us too far away from the main purpose of this book.

Let us now turn to a problem that belongs to the history of science. Another Nobel Prize winner, the American theorist Stephen Weinberg wrote in his book *The First Three Minutes: A Modern View of the Origin of the Universe*:

> I want especially to grapple here with a historical problem that I find both puzzling and fascinating. The detection of the cosmic microwave radiation background in 1965 was one of the most important scientific discoveries of the twentieth century. Why did it have to be made by accident? Or to put it another way, why was there no systematic search for this radiation, years before 1965?.

Can the question be answered by pointing out that scientists did not have sufficiently sensitive radio-telescopes capable of detecting the microwave background? We can see that this explanation is unlikely. Weinberg was also of this opinion. This is not the most important factor, however.

We can find a considerable number of examples in the history of science when the prediction of a new phenomenon was made long before experimental confirmation became technically feasible. Nevertheless, if the prediction was important and had solid foundation, physicists never lost it from view. It was invariably tested once proper equipment became available. Weinberg cites as an example the prediction of the antiproton, made in the 1930s (the antiproton is the antiparticle of the nucleus of hydrogen). Experimental detection was out of the question at the time. In this case, however, twenty years later the situation had changed and a special accelerator was constructed in Berkeley to test this prediction. However, astronomers did not even know about the predicted microwave background or the possibility of its detection.

Why did this happen? Weinberg gave three reasons. First, the hot universe theory was developed by Georgy Gamow and his colleagues to explain the abundances of *all* the chemical elements by their synthesis at the very beginning of the expansion of the universe. This assumption proved to be erroneous. We have already mentioned that heavy elements are synthesised in stars. Only the lightest elements originate from the first moments of expansion. The first versions of the theory contained other faults too. All necessary corrections were later made but in the 1940s and 1950s the theory was not regarded as plausible.

Weak links between theorists and experimentalists constituted the second reason. Theorists did not think that the microwave background could be detected by available instruments, while experimentalists never heard that it would be wise to search for such radiation.

Finally, the third reason was psychological. Physicists and astrophysicists had to overcome a barrier preventing them from starting to think that calculations dealing with the first minutes after the Big Bang could really correspond to the true situation. Indeed, the contrast between the time intervals was too great: the first several minutes – and the fifteen billion years separating our time from time zero.

Another reason, which seems to be the most important to the authors of this book, was characterised by A. Penzias in his Nobel lecture. This was that neither the papers of Gamow and his colleagues nor later publications about the primordial radiation mentioned that the background could be detected, even in principle. Moreover, it appears that Gamow and his colleagues thought that this was impossible in principle! Penzias said:

As for the detection, they appear to have considered the radiation to manifest itself primarily as an increased energy. This contribution to the total energy flux incident upon the Earth would be masked by cosmic rays and integrated starlight, both of which have comparable energy densities. The view that the effects of three components of approximately equal additive energies could not be separated may be found in a letter by Gamow written in 1948 to Alpher *[unpublished and kindly made available to A. Penzias by R. A. Alpher from his files]*. 'The space temperature of about 5 K is explained by the present radiation of stars (C-cycles). The only thing we can tell is that the residual temperature from the original heat of the Universe is not higher than 5 K'. They do not seem to have recognized that the unique spectral characteristics of the relic radiation would set it apart from the other effects.

At the beginning of the 1960s, A. G. Doroshkevich and one of the authors of this book (I. Novikov) published a paper in which they demonstrated that even though the total amount of energy in the primordial radiation is comparable with the energy coming from galaxies (taking into account their evolution and the expansion of the universe), the background radiation is concentrated in the range of centimetre and millimetre wavelengths, where both galaxies and ordinary radio sources emit only very weakly. Hence, the background can be observed!

We will quote Arno Penzias again:

The first public recognition of the relict radiation as a detectable microwave phenomenon appeared in a brief paper entitled *Mean Density of Radiation in the Metagalaxy and Certain Problems in Relativistic Cosmology* by A. G. Doroshkevich and I. D. Novikov in the spring of 1964. Although the English translation appeared later the same year in the widely circulated *Soviet Physics – Doklady*, it appears to have escaped the notice of other workers in this field. This remarkable paper not only points out the spectrum of the relict radiation as a blackbody microwave phenomenon, but also explicitly focuses upon the Bell Laboratories 20-ft horn reflector at Crawford Hill as the best available instrument for its detection!

This paper remained unnoticed by both theorists and observational astronomers until the microwave background was accidentally discovered; it did not lead to intentional search.

Incidentally, the microwave background could have been discovered as early as 1941! At that time, the Canadian astronomer E. McKellar analysed absorption lines caused in the spectrum of ζ Ophiuchus by interstellar cyanide molecules (HCN). MacKellar concluded that these lines in the visible part of the spectrum can only be explained as resulting from the absorption of light by rotating cyanide molecules. The rotation of molecules was excited, he concluded, by a radiation with a temperature of about 2.3 K. Neither McKellar himself nor anyone else even thought that the rotational levels of molecules could be excited by the microwave background. Even the theory of the hot universe had not yet been developed!

Only after the microwave background had been discovered did I. S. Shklovsky, J. Fild, J. Hitchcock, P. Thaddeus and J. Wolf publish papers showing that the primordial radiation did indeed excite the rotation of interstellar cyanide molecules observed in the spectra of ζ Ophiuchus and other stars. We thus see that evidence of the microwave background, even though indirect, had already been observed in 1941.

This is not the end (or rather the starting point) of the story, however.

Let us return to the problem of the feasibility of detecting the primordial radiation. The question is: when did the equipment become capable of achieving this? Steven Weinberg wrote:

It is difficult to be precise about this, but my experimental colleagues tell me that the observation could have been made long before 1965, probably in the mid-1950s and perhaps even in the mid-1940s'.

Was this indeed possible?

In the mid-1950s, a young Soviet astrophysicist, T. A. Shmaonov, working in the group of the well-known astronomers S. E. Khaikin and N. L. Kaidanovsky, measured the cosmic radiation at 3.2 cm wavelength. These measurements were conducted with a horn antenna, very similar to the antenna which much later was used by Penzias and Wilson. Shmaonov had carefully analysed possible sources of noise. Obviously, he had nothing comparable to the high-sensitivity receivers available to the American group. Shmaonov's results were published in 1957 in his PhD thesis and in the Soviet journal *Pribory i Teknika Eksperimenta*. The conclusion he drew from the measurement results was: 'The absolute effective temperature of the background centimetre radiation ... was 4 ± 3 K'. Shmaonov emphasised that the radiation intensity was independent of time and of the direction in the sky. Even though Shmaonov's measurement errors were high and the figure he gave (4 K) was quite unreliable, we now understand that it was the primordial radiation that he had found. Unfortunately, neither Shmaonov nor his senior colleagues, nor other radioastronomers, knew anything about the microwave background and consequently they attached no importance to Shmaonov's data.

This is the complex knot of events that preceded the discovery of the hot universe by Penzias and Wilson. The recognition that the universe at the initial moment was heated to a superhigh temperature was the starting point for some extremely important research, leading to the uncovering of some of the mysteries not only of astrophysics but also of the structure of matter. The scienctific schools of thought created by V. L. Ginzburg, S. Hawking, I. M. Khalatnikov, A. M. Markov, J. Ostriker, P. J. E. Peebles, M. Rees, S. Weinberg, Ya. B. Zeldovich, A. L. Zelmanov and others made important contributions to the development of various aspects of modern theoretical cosmology.

The latest discoveries in this field are discussed in the final section of this book.

Explosion

Hubble's discovery of the expanding universe formulated a question of tremendous philosophical importance: how had the universe exploded?

Friedmann's theory describes how this expansion evolves under the action of gravitational forces. Galaxies recede from one another, moving by inertia, and the gravitational forces gradually slow down their motion and with it the expansion of the universe.

However, the theory does not answer the question of what triggered the expansion. What imparted the initial velocities to the matter from which galaxies were later formed?

The discovery of the microwave background demonstrated that the universe at its birth was hot, and the pressure of matter, which was almost uniformly distributed in space, was tremendously high.

At first glance, the high pressure was a very important factor. We can recall the explosion of an explosive charge, releasing energy in a small volume. This can be chemical or, say, nuclear energy. The matter immediately heats up and evaporates. The pressure of the hot gas drives its rapid expansion. When we think of the origin of the universe, the picture outlined above invites itself. However, are high temperature and high pressure indeed the reasons causing the expansion? No, they are not. There is a substantial difference between the two phenomena. In the explosion of a charge, we deal with a pressure drop: very high pressure of hot gas in the explosive and relatively low atmospheric pressure outside it (if the explosion occurs in the air). This pressure difference creates the forces that hurl the matter apart. It is the pressure drop, not the high pressure in itself, that produces these forces. Obviously, matter would not fly apart if the pressure of the exploding gas were the same inside and outside the explosive. Furthermore, the density of the expanding hot gas is non-uniform in the process of explosion: it is maximal at the centre and falls off towards the boundaries. As the gases expand, the pressure drop, related to the drops in density and temperature, produces the force pushing out the expanding gas.

When the universe starts to expand, the picture is very different. Its matter was uniform before heavenly bodies formed and no density and pressure differences existed. No force was generated, therefore, to prime the expansion. Hence, high pressure of hot gas is not the cause of the expansion of the universe.

What, then, was the 'pre-push' that imparted initial velocities to matter? In order to reconstruct the processes that occurred at the very start of the expansion, we need to find traces of these 'most ancient' processes in today's universe.

It was found that such 'traces' are the fundamental properties of the universe in which we exist. They look mysterious if we do not succeed in explaining the origin and evolution of these properties in the course of the explosion of the universe.

The first of these puzzles is the uniformity of the universe on a large scale. Observations show that, on a scale greater than several hundreds of megaparsecs, clusters of galaxies are indeed distributed uniformly. Actually, it is very difficult to arrive at reliable conclusions for such distances, because of the difficulties in observing very distant faint objects. Such observations do not give direct data about the distribution of the 'dark matter', that is, invisible (non-luminous) forms of matter. Nevertheless, the conclusion of large-scale uniformity of the universe both for visible and (which is especially important) for invisible matter is quite reliable. How was it obtained?

The tool for the investigation was the microwave background radiation.

Today, the universe is completely transparent for this radiation, but it was not so in the distant past. When the temperature was above 4000 K, the entirety of matter existed in the form of ionised plasma (there were no heavenly bodies at that stage), which is opaque for the microwave background radiation. The transformation of plasma into neutral matter occurred some 300 000 years after the expansion started; beginning with this epoch, most primordial photons moved along straight lines, never interacting with neutral atoms. Therefore, when we observe the microwave background, we look into the past, into that distant time which we call the recombination epoch, because it was at this time that electrons were captured by atomic nuclei and neutral matter was formed. In the period since the recombination epoch, the radiation traversed about 15 billion light years. This is the maximal distance that light travels in the expanding universe even if it had been emitted at the very beginning of the expansion, that is, 15 billion years ago. Correspondingly, this distance is called the distance to the apparent horizon. Therefore, our observation of the microwave background 'scans' practically the whole region of the universe accessible to observational instruments.

What is the role of the microwave background in solving the problem of the degree of uniformity of the universe? The key fact is that this radiation carries information about the properties of the universe at points which lie at great distances from one another. These properties are found to be surprisingly similar. For instance, measurements performed with the RELIKT satellite of the Institute of Space Research (Moscow), with the American Cosmic Background Explorer (COBE) satellite and with other instruments demonstrated that the intensity of the microwave background coming to us from diametrically opposite points in the sky coincides to an accuracy of at least several hundredths of one percent. Since each such ray of radiation arrives with us from the horizon, the areas from which the microwave background was emitted are now separated by 30 billion light years. The uniformity of the radiation is evidence that the properties of these areas are identical.

However, what is so surprising in the fact that the universe is uniform on a large scale? The reason it is surprising is the following. A light signal emitted from one of the points even 15 billion years ago cannot travel a distance of 30 billion light years. Nothing can move faster than light. Hence, no signal was able to reach a point at a distance of 30 billion light years from another. There is thus no reason for equalising or 'matching' the conditions at these points, because there was no possibility for these points even to exchange signals. Nevertheless, the conditions at these points are identical. Why?

This is the first puzzle which the theory must solve. It is known as the 'horizon problem'.

Let us now look at the second fundamental property of the Universe, which also requires explanation. We have already said that the expansion of the universe is slowed down by gravitational forces. These forces, determining the energy of gravitation, depend on the mean density of matter in the universe. At the same time, the velocities of recession of galaxies from one another determine the kinetic energy of the expansion. If the energy of gravitation at the beginning of the expansion was much greater than the initial kinetic energy of recession, the expansion would have stopped long ago and the universe would now be contracting. On the other hand, if the kinetic energy at the beginning was much higher, galaxies would today fly apart by inertia, with no deceleration caused by gravitation. The value of the density of matter at which both these energies are equal is called the critical density. Observations show that the density at the first moments of expansion was extremely close to the critical density. As an illustration, we can consider a moment of time in the past very close to the start of expansion, when, according to current theory, the unified physical interaction decayed and the strong nuclear interaction began to play an independent role. This moment is known as the epoch of 'Grand Unification'; it is separated from 'time zero' by only 10^{-33} s. According to observational data on today's recession velocity and mean matter density, and according to the Friedmann model, the deviation of the density from the critical value at the 'Grand Unification' epoch was less than 10^{-50} of the value of that density itself!

Therefore, the matter density in the universe at the very beginning of the expansion was unimaginably close to critical. But why? Why did nature tune the power of the explosion so finely that the real density of matter in the universe coincided to a fantastic accuracy with the critical value?

This is the second mystery of the universe, sometimes known as the 'critical density problem'.

Another problem is that despite the surprising uniformity of the universe on a very large scale, something caused deviations from uniformity

– small initial fluctuations – on smaller scales. What were these factors? It was these fluctuations that caused, under the action of gravitational forces, the formation of small density peaks that evolved and formed, in an epoch closer to ours, both galaxies and clusters of galaxies.

Finally, there is another problem. It is connected with very unusual particles predicted by current theories, for example, magnetic monopoles. These unusual particles would have been created in the universe during the 'Grand Unification' epoch. They should have been created in great abundance. Actually, some fraction of these monopoles annihilated with their antiparticles (antimonopoles) in the course of further evolution. Calculations by Ya. B. Zeldovich and M. Yu. Khlopov demonstrated that today's universe must abound in monopoles, containing about as many of them as it has protons. However, monopoles are more massive than protons by a factor of 10^{16}. Hence, the density of monopole matter in today's universe would be 10^{16} times greater than that of ordinary visible matter. This scenario is definitely out of the question. We conclude that there are practically no monopoles in our universe. How did they disappear?

This mystery is knows as the 'monopole problem'.

All these puzzles are connected with processes that took place at the very beginning of the expansion of the universe, that is, these processes hide the mystery of the origin of the universe in an encrypted form. It was necessary to discover the key to the cipher.

In what follows, we will outline the hypotheses which describe, according to current notions, the first moments of the Big Bang. The key to understanding the 'primary pulse' lies in the creation of a very special, so-called vacuum-like state of matter which can be created when the matter density is very high. Modern physics defines very high density as the density close to the quantity determined by three fundamental constants, the gravitational constant G, Planck's constant h and the velocity of light c:

$$\rho_{Pl} = \frac{c^5}{G^2 h} \approx 10^{94} \, \text{g/cm}^3.$$

The immensity of this quantity is difficult to comprehend. This density is known as the Planck density. According to the theory, unusual states characterised by gigantic tensions, or, which amounts to the same thing, by negative pressures, may arise in matter. The relation of the density ρ_* and the pressure P_* of such states is $P_* = -\rho_* c^2$. These are states known as *vacuum-like states*.

The origin of this term is connected with the following fact. Even if all real particles and fields were removed from some area of today's Universe, this region could not be treated as 'absolutely empty space' (the vacuum). The fact is that the vacuum is a place where so-called virtual pairs (particles and antiparticles) are constantly created and annihilated, where very unusual 'quantum fluctuations of the vacuum' occur. The corollaries of these processes have been measured in elaborate experiments. Quantum fluctuations of the vacuum cannot be suppressed. One possible consequence of these processes is the presence of very low vacuum density ρ_{vac} and negative pressure p_{vac} (physically, negative pressure signifies tension). These quantities obey the relation $p_{vac} = -\rho_{vac}c^2$. Any state in which pressure and density obey this relation is called a vacuum-like state. The specific feature of such states is that they remain unchanged by expansion: the density and pressure stay constant.

Another important factor is the modification of Newton's law of gravitation in Einstein's theory. According to Einstein, gravitational accelerations are created not only by the mass density ρ but also by pressure P (or tension). Instead of ρ, the formula for gravitation includes the sum $\left(\rho + 3P/c^2\right)$. Under ordinary astrophysical conditions, the second term is extremely small. However, it becomes decisive in the case of the vacuum-like state. Substituting then $P_* = -\rho_* c^2$, we find that the sum in parentheses becomes negative and gravitational attraction changes to repulsion. This repulsion is not hydrodynamic (as in the case of a pressure drop) but of a purely gravitational character; in all likelihood, this repulsion served as the 'primer' for the expansion of the universe.

Any two particles in this very early universe moved at accelerating velocity relative to one another. The density ρ_* of the vacuum-like state, as we have already stated, was not decreasing in the course of the expansion, neither did the tension (negative pressure) P_* decrease; the accelerating force was constant.* It is not difficult to show that in this case distances between particles increase exponentially, that is, extremely rapidly: $R = R_0 \exp(3 \times 10^{43} t_{(s)})$. This process is known as *inflation*.

Presumably, this stage lasted from the time $t \approx 3 \times 10^{-44}$ s, when the mass density both of particles and of the vacuum-like state were close to Planck's value $\rho_{Pl} \approx 10^{94}$ g/cm^3, up to the moment $t \approx 3 \times 10^{-35}$ s. By the end of this period, all particles had receded to unimaginably large

---

* This accelerating force is described in a different language by introducing the already mentioned Λ term into Einstein's equations. This aspect was pointed out by Ya. B. Zeldovich.

distances: about $10^{4 \times 10^8}$ parsec from one another. For a comparison, the reader may recall that the size of the universe that we observe today is a 'mere' 10^{10} parsec! In this early universe, there were virtually no particles (they were extremely rare) and the temperature was practically equal to zero. The only thing left in the universe when inflation had been completed was the vacuum-like state. However, this state was unstable; at t approximately equal to 3×10^{-35} s, it decayed to ordinary particles moving with ultrarelativistic velocities. The temperature in the universe in the course of the decay of the vacuum-like state jumped to $T \approx 10^{27}$ K. The universe became hot! This was the end of inflation: the vacuum-like state had disappeared. Further expansion of the universe began to slow down, owing to the gravitational interaction between particles of ordinary matter. The subsequent fate of the expanding hot matter was described in the preceding section.

The assumption that tremendous negative pressure and, hence, gravitational repulsion may appear at a very high density of matter was formulated at the end of the 1960s by E. B. Gliner. In 1972, D. A. Kirzhnits and A. D. Linde showed that such a state can be naturally formed in an expanding universe in which the temperature and density fall off from very high values. Somewhat later, these pioneering ideas were extended to cosmology in papers by E. B. Gliner, L. E. Gurevich and I. G. Dymnikova. After this, A. Albrecht, A. Guth and P. Steinhardt in the USA and A. D. Linde and A. A. Starobinsky in the USSR, and several others, continued this work, using the latest results of high-energy physics.

The inflation of the universe is the key to solving the mystery of its fundamental properties.

Let us begin with the first problem: the horizon problem. The difficulty is that sufficiently remote points could not exchange light signals even until today, so that one point could not 'be aware' of conditions at another point. It is not clear, therefore, why the temperatures and other observed physical parameters at these points could become identical. Inflation provides the following explanation. Indeed, points which are so remote in today's universe cannot exchange signals, but they could do so only in the initial universe in which there was no exponential inflation. Inflation explodes the distance between any two points. Therefore points which are remote today were very close at the start of inflation: they were inside a region of about 10^{-33} cm in diameter, that is, these points practically coincided and the exchange of signals between them was unobstructed. There is thus nothing surprising in the identity of conditions at points which receded from practically the same 'point'.

Puzzle two: the unknown reason for the minute difference between the current density in the universe and the critical density, and for the practical equality of the two densities in the remote past, especially at the origin ('time zero').

Inflation solves this problem in the following way. The acceleration produced by gravitational repulsion imparts a kinetic energy to the exploding matter exactly equal to the gravitational energy. When the vacuum-like state decays at the end of the inflation stage and transforms into ordinary matter, the density ρ_* transforms into the density of ordinary matter ρ; it is not surprising that the energies of gravitation and of outward motion are exactly equal and that the density equals the critical value.

Puzzle three: the origin of small primary fluctuations in the matter density, which later gave rise to galaxies and their star systems. The point is that the decay of the vacuum-like state is a quantum process subject to random fluctuations typical of such processes (for instance, radioactive decay). At some points the decay of the vacuum-like state could occur, for accidental reasons, a little earlier than at other points and, hence, would cause there the transition to the hot universe. Calculations show that this factor generates small-amplitude fluctuations in the density of the emerging hot matter. A large contribution to the solution of this problem was made by J. Bardeen, G. S. Chibisov, A. Guth, S. Hawking, A. D. Linde, V. N. Lukash, V. M. Mukhanov, A. A. Starobinsky, P. Steinhardt, M. S. Turner, A. Vilenkin and many others.

Finally, the fourth puzzle is that of the monopoles. Its solution in terms of inflation is obvious. Monopoles are created in the universe at the first moments of inflation when the temperature is still very high. After this, the inflation of the universe blows monopoles apart to gigantic distances. They are so rare now that it is almost impossible to come across them in the universe.

This is a schematic outline of phenomena which are likely to have taken place at tremendously high matter densities and energies; they led to the 'primer push' and later, after a long chain of events, to the universe we observe today. What, however, was there even earlier?

It is not easy to answer this question, and not only because experts know precious little about processes at the Planck density and energy. The difficulties arise because the fundamental features of such general categories of the existence of matter as space and time get drastically modified under these extreme conditions.

It is probable that the Planck density is the maximum density allowed in nature. Space and time at such densities break down into 'quanta' characterised by time intervals $t_* \approx 3 \times 10^{-44}$ s or spatial extensions $r_* \approx t_* c \approx 10^{-33}$ cm. On such scales, the 'fluctuations' of the vacuum are very violent, and time intervals less than t_* are meaningless. This state of matter is known as the singularity. All the properties of the universe that we observe today actually arise in this singular state. This leads to another question.

Einstein once said that the aspect of profound interest to him was whether God could create the world to be different from the one around us. When the great physicist referred to 'God', he meant nature, so that we can interpret his words as a question of whether the universe could be arranged in a different manner. Some time ago, a physicist would shun such questions; nowadays this is a completely legitimate field of research for modern physics and astronomy.

The problem can be formulated in the following form. What would happen if the laws of physics were different? For instance, what if the charge of the electron (or proton) were changed several-fold? Or what if the electron mass were different? Here is one possible answer. The value of the particle charge determines the attractive force between a proton and an electron, and the electron mass dictates the behaviour of electrons in bound states in atoms. Hence, modifications like this would change the size of atoms and, consequently, the dimensions of the bodies that surround us. If the changes in the properties of the electron were small, the alterations in the surrounding bodies would also be minor.

A similar answer seems to be logical if we want to predict the consequences of changes in the gravitational constant G. Obviously, the force of the gravitational interaction of the same masses would change. This force determines the rate of evolution of heavenly bodies and their sizes. All these characteristics would be modified. Again, the variations in the properties of heavenly bodies can be expected to be small if the changes in the gravitational constant are not large.

It also seems plausible that similar answers would be obtained to questions about changes in other physical constants. The general answer seems to be that the *Gedankenexperiment* of relatively small variation of the physical constants would result in corresponding small quantitative changes in the surrounding world. No qualitative, profound modifications are thus anticipated in the Universe in response to such variations.

An analysis shows that this conclusion is completely wrong.

As an example, we can consider the simplest system: the hydrogen atom. This atom can exist for an indefinitely long time if it is not subjected to external factors. The electron and the proton in a neutral atom do not enter into a reaction creating a neutron and a neutrino, even though there is a finite probability for the electron to be at the point of location of the proton. However, this reaction does occur when a high-energy electron collides with a proton. The impossibility of the reaction in a neutral atom stems from insufficiently high energy. The sum of the electron and proton masses is less than the neutron mass. The difference is $E \approx 0.8$ MeV. Hence, it is impossible to make a neutron out of a proton and an electron without expending at least these 0.8 MeV. If we imagine that the electron mass is not 0.5 MeV but exceeds the mass difference of the neutron and the proton, $\Delta m = 1.3$ MeV, the neutron-forming reaction becomes possible. For example, if the electron mass were 2 MeV, the neutral hydrogen atom would only exist for 30 hours. Therefore, the hydrogen atom can exist for very long only if the inequality $m_e < \Delta m$ holds. The same result is obtained if the electron mass is invariant but Δm is reduced to values much below m_e. We need to emphasise that the reduction of Δm needed to allow the neutron-forming reaction in a hydrogen atom (Δm must be lowered by only 0.8 MeV) is absolutely negligible in comparison with the total mass of the proton or neutron (both of the order of 1000 MeV). A change in the mass of these particles by about 10^{-3} of their value would result in catastrophic consequences: the absence of hydrogen in today's universe. This would mean no conventional nuclear fuel for stars. The tiniest variations in the masses of elementary particles would remove the stars of the main sequence from the universe; there would be no hydrogen-containing chemical compounds, and life in such a universe would most likely be impossible.

Hence, small variations in the above parameters would lead not to small changes in the properties of heavenly bodies but to qualitative changes in the properties of the universe.

The example given above is the rule rather than an exception.

To support this statement, consider the properties of heavy hydrogen, deuterium. The binding energy of particles in the atomic nucleus (deuteron) of this element is $\mathscr{E}_b = 2.2$ MeV. The fact that the energy $\mathscr{E}_b$ is greater than $E = 0.8$ MeV means that the deuteron is stable. It is 'energetically unfavourable' for the neutron in the nucleus to decay into a proton, an electron and an antineutrino and thus destroy the deuteron. Hence, the necessary condition for the stability of deuterium is $\mathscr{E}_b > E$, which we can rewrite in the form $\Delta m < \mathscr{E}_b + m_e$. What would be the consequence of the violation of this inequality, and with it, the instability of the deuteron? Even though the amount of deuterium in nature is quite small (about one hundred thousandth of the entire mass of the universe), it plays a very prominent role. The deuteron is the first complex atomic nucleus in the chain of nuclear reactions resulting in the formation of heavier nuclei. Such reactions occurred also at the very beginning of the expansion of the universe, and continue to occur inside stars, in the processes converting hydrogen into helium. If there were no deuterium, the standard mechanism of forming elements heavier than hydrogen would be impossible. It would again lead to drastic qualitative changes in the universe. Altogether, two inequalities must hold simultaneously to make the 'properties of the universe stable': $m_e < \Delta m < \mathscr{E}_b + m_e$, which is a rather fine tuning of the fundamental physical constants.

Continuing this line of argument, we can add the following fact. The constant of the strong interaction is such that the nuclear forces are sufficient for holding the neutrons and protons together in complex atomic nuclei. If this constant were slightly lower, the nuclear forces would be insufficient for the stable existence of all complex atomic nuclei. This means that chemical elements heavier than hydrogen would never form in nature. Neither nuclear processes in stars, nor chemical forms of matter, nor life itself would be possible.

Finally, let us consider another corollary, this time implied by an imaginary variation of the gravitational constant.

We know that if a star is of approximately solar mass or less massive, its surface layers of considerable thickness undergo convective mixing. In contrast to this, more massive stars have no surface convective layers after birth. There is a hypothesis that the formation of planetary systems, which takes place together with the formation of stars, is possible only around those stars which retain their surface convection.

An analysis showed that if the gravitational constant were considerably higher than we find today, no stars would have convective surface layers after formation and, hence, would probably have no planetary systems. In all likelihood, life in such a universe would be impossible. Even though these arguments are based on a number of hypotheses, the conclusion is quite impressive.

We will not give further examples and only remark that careful analysis points to the following conclusions. Variation of some constants may result in the impossibility of formation of galaxies, stars or even elementary particles! This means that no complex structures could evolve in the universe.

Therefore, relatively small variations in fundamental constants result not only in small quantitative alterations but in cardinal qualitative changes in the universe. In this sense, the universe is found to be highly unstable with respect to such variations of the laws of physics.

The question now is: are such arguments meaningful at all? 'Who' or 'what' can vary the laws of physics when in reality we do know the values of the actual fundamental constants and, using these values, must analyse all the processes in nature. Indeed, do we suspect that there exist some 'other' physics or 'other' universes?

Before considering attempts to answer these questions, let us look at the following amazing facts belonging to 'our' physics and 'our' universe. Sometimes, the conditions that we have enumerated above, and other conditions necessary for the existence of complex structures, are indeed perceived as quite strange.

Indeed, let us turn to the first condition written as the inequality $m_e < \Delta m$. According to this inequality, the electron mass must be small and not only small but less than 1.3 MeV. Let us now look at the list of elementary particles. The electron is the lightest of particles with non-zero mass. It is two hundred times lighter than the next-lightest particle, the muon. The spectacular feature is that the electron is not simply lighter than other particles but that it is much lighter. The masses of almost all other 'ordinary' elementary particles do not differ too much, being all of the order of 1 GeV. The electron is in clear opposition, deviating to much smaller masses against this background. The special significance of these facts was pointed out by I. L. Rozental.

The very low mass of the electron looks like a giant fluctuation. If there were no such fluctuation and the electron were, say, only several times lighter than the muon, then the inequality $m_e < \Delta m$ would be violated, with all the catastrophic consequences it implies.

Let us turn now to the inequality $\Delta m < \mathscr{E}_b + m_e$. It demands that the difference Δm of the neutron and proton masses be sufficiently small.

The proton and the neutron are similar particles that differ only in their charges and small mass difference. Such families of similar particles are known as 'isotopic spin multiplets'. If we look at a list of mass differences in other such families that are stable with respect to the strong interaction between particles, we find that Δm for the neutron and the proton is the smallest of all. Another fluctuation! And again precisely the one required to support the inequality needed for the existence of complex structures.

These examples show that the values of the constants often create an impression that nature intentionally 'adjusted' these values to ensure the development of complex structures in the universe and, finally, the appearance of life. Nature is sometimes 'forced' to arrange large fluctuations from the typical values of constants, to resort to extreme fine tuning of the laws of physics.

Note that our universe reveals one more sign of 'curious behaviour'. We mean that the lifetime of a typical star coincides, by an order of magnitude, with the time elapsed since the universe began to expand.

This coincidence does look very strange. It can be shown that a star's lifetime is determined by the rate of nuclear reactions in it and by the opaqueness of its matter, that is, it is dictated in the final analysis by the properties of protons and electrons, and by the force of the gravitational interaction which is determined by the gravitational constant G. On the other hand, the 'lifetime of the universe' is a function of quite different processes, that is, the processes taking place at the very beginning of the Big Bang.

What, then, is the message contained in the closeness of these times – is it a mere coincidence or is it very important?

A new scientific approach has been formulated relatively recently, attempting to answer this question and also to explain all the puzzling features and 'strangeness' in our universe (we have already mentioned some of them above). A well-known Soviet cosmologist, A. L. Zelmanov, said, characterising this approach: 'We are witnessing these events of the past simply because other events take place without witnesses'. These words express the essence of the so-called *anthropic principle*.

What is this principle and what is its relation to the problems we discuss here?

Note, first of all, that complex forms of motion of matter, such as complex chemical compounds, life and, even more so, intelligent life, could arise in the universe only at a specific stage of its evolution, fairly close to our epoch. Indeed, complex chemistry and life phenomena require, at least in the forms we are aware of, the existence of Earth-like planets, with oceans heated by a star at a sufficiently close distance, whose luminosity remains constant for a long time. Of course, life can arise only on the basis of complex chemistry and will take many billions of years to evolve.

Such conditions were definitely absent in the distant past of the universe when it had neither stars nor planets. Neither can life originate in the distant future when the stars burn out, and even less so in that very distant future when the heavy particles decay into light and neutrinos.

The first conclusion then is that life and specifically intelligent life of the type we witness on Earth can appear in the universe only at a very specific period: in our epoch, when conditions are suitable for it.

The anthropic principle thus explains the seemingly strange coincidence of the lifetimes of stars and of the age of the universe. The explanation is this: for intelligent life to appear in the universe, the age of the universe must be approximately equal to the age of a star; this looked puzzling at first glance.

Another conclusion from the anthropic principle is that observers ('witnesses') can appear only under a very specific set of physical constants, under quite definite physical laws; we have mentioned this above. If there were (or, perhaps, are?) other universes with different laws, they existed (or exist) without complex structures and, hence, without 'witnesses'. Life can *never* appear in such universes. Therefore, our universe is the way it is and the way we see it precisely because we exist in it.

The anthropic principle was actively elaborated and is still worked on by such well-known physicists and astronomers as John D. Barrow, Brandon Carter, Robert Dicke, Paul Dirac, Georgy Gamov, Stephen Hawking, Martin Rees, Iosif L. Rozental, Joseph Silk, John A. Wheeler, Yakov B. Zeldovich, and others.

Another fundamental feature of our world is the fact that physical space is for some reason three-dimensional; it is not two-dimensional, or five-dimensional but precisely three-dimensional. Physicists recognised fairly long ago that there was a puzzle there. Ernst Mach formulated the question unambiguously: 'Why is space three-dimensional?' Paul Ehrenfest was the first to begin a serious analysis of the problem.

To try to comprehend the essence of this problem, we can follow the recipe already used in thinking about other fundamental constants: imagine that the number of spatial dimensions has been changed, that is, try to imagine the consequences of a dimensionality different from three.

We will spend some time now considering only several modifications in the simplest physical interactions, caused by the variations of the dimensionality of space.

One of the simplest examples of physical interactions is Coulomb's law for charges at rest and Newton's law for gravitating masses. In both cases, the interaction force decreases in inverse proportion to the distance squared, $F \propto 1/r^2$. However, Immanuel Kant was already aware that the inverse square law is a corollary of the three-dimensionality of our world. Indeed, why does the force of, say, electrostatic interaction decrease with distance? The most obvious answer is that as r increases, the lines of force of the field are spread over a gradually greater surface area of the sphere of radius r enveloping the charge. The area of the sphere grows in proportion to r^2, so the density of the lines of force, piercing this sphere, decreases in proportion to r^{-2}, which dictates the law of variation of the force. However, all this holds only in three-dimensional space. If space is four-dimensional, the surface area of a three-dimensional sphere (the locus of points equidistant from the centre in four-dimensional space) is proportional to r^3, and in a space of dimensionality N, this surface area is proportional to r^{N-1}. Hence, the law of variation of the electrostatic and gravitational forces in N-dimensional space is $F \propto 1/r^{N-1}$. Why is it so important to measure the dependence of force on distance as it decreases with r in an N-dimensional space? Let us consider the motion of a trial charge on a circular orbit around the central charged body (we assume it to be of the opposite sign of charge, to ensure an attractive force) in a space of arbitrary dimension N. Let the orbital momentum of the charge be fixed (it cannot vary during motion). The centripetal force is then proportional to r^3 and does not depend on N. We know from mechanics that for *stable* circular orbits to exist, it is necessary that the centripetal force decrease with distance faster than F does. Otherwise the circular motion becomes unstable and the tiniest perturbation either causes the charge to fall into the centre or to recede to infinity. In its turn, the absence of a stable circular orbit signifies the absence of any bound state for a charge moving in a restricted region of space around the central body. This means that for bound states, we need $N \leq 3$. This result was later extended to quantum mechanics.

This result is rather unexpected. It may seem, at first glance, that an increase in the dimensionality of space would open new possibilities for more complex modes of motion of bodies in this space, and hence, for structurally more complex systems. Actually, we find that such spaces would have no stable bound systems of bodies interacting through electric or gravitational forces, that is, they cannot have either atoms or planetary system, or stars, or galaxies!

On the other hand, if $N = 2$ or 1, interacting charges of opposite sign would never be able to recede to infinitely large distances in such spaces. Forces here decrease too slowly; whatever the initial velocity imparted to a charge, the central body stops the receding charge and forces it to move centrewards. Such spaces do not allow the free motion of gravitating bodies.

The existence of both bound and free states is possible only in three-dimensional space.

After these words, it is not going to look too strange if we say that if nature had to try many times to 'create' universes with different spatial dimensionalities, the possibility of the existence of both bound gravitating systems and free bodies and of the existence of bound and free states of motion of electrons in atoms would arise only if $N = 3$. Hence, complex and diverse structures possessing the ability to arise and decay can evolve only in this scenario. This is the only scenario allowing variability, evolution and the birth of life; hence, 'witnesses' can exist precisely in such worlds (perhaps only in such worlds!). It is not surprising, therefore, that we live in a three-dimensional space.

In fact, modern theories of physical processes at superhigh energies, which aim at merging all forces in nature, state that perhaps we live not in three-dimensional space but in a space of a much greater number of dimensions, say, nine-dimensional space. The additional six dimensions are 'compactified': twisted to unimaginably small scale, so that it is absolutely impossible to move in these directions. Even if this picture is true, only three 'effectively present' dimensions are observable.

In order to solve the above-mentioned problems of the 'strange' properties of our universe, one has 'only' to find out whether nature had indeed 'tried to create' numerous universes, or even an infinitely large number of them with different physics and sometimes with fluctuations of the numerical values of constants, with a different dimensionality of space, and so forth. It would then be clear that we – the observers, the researchers – did appear only in the rarest of the universes, the 'most fortunate' of them (fortunate for our existence).

The American physicist John Archibald Wheeler has constantly emphasised for the last thirty years the principal importance of the quantum fluctuations of the properties of space–time; they are expected to take place at the Planck density $\rho_{Pl} \approx 10^{94}$ g/cm^3, on a scale of about $r_* \approx 10^{-33}$ cm and $t_* \approx 3 \times 10^{-44}$ s. Here space–time is, in a certain sense, a 'breathing' foam of black and white holes which arise and immediately disappear, of very tiny mini-universes and of even more complex topological structures. A. D. Linde and A. A. Starobinsky were able to develop and elaborate these concepts in terms of modern physics and cosmology.

According to the picture outlined by Linde, the predominant part of physical space–time exists in the state of quantum foam, at a density close to $\rho \approx 10^{94}$ g/cm^3. Quantum fluctuations occur in 'bubbles' created in the foam; the gravitational repulsion in the vacuum-like state then inflates these bubbles. The predominant part of the volume of the 'bubbles' immediately returns to the 'foam' state via new fluctuations. In a smaller part, inflation may continue and quantum fluctuations of the density of the vacuum-like state may appear. A very small fraction of the initial volume may, after a long chain of random fluctuations, have a vacuum-like state density considerably less than ρ_{Pl}. At this stage the amplitude of quantum fluctuations is not as high as it was before. These volumes continue to inflate, as described at the beginning of this section, and transform into hot universes after the vacuum-like state decays.

We are in one such universe. It can be said that the universe (or, if you prefer, multiple universes) is being permanently reborn from fluctuations, that the universe is always reproducing itself. This world taken in its entirety has no initial point and will never end. This is a picture of exploding eternity.

When new mini-universes are born from the vacuum foam, it is likely that all physical parameters, including the dimensionality of space and time, and the physical laws themselves undergo fluctuations. Nature has thus 'tried' an infinite number of times to create universes with most diverse properties. We live in the 'most favourable' (for us) sample of this eternal creative process. We should not forget, however, that 'our universe' is neither the most typical nor the most probable part of the totality.

This is the answer that modern science gives to Einstein's question about the possibility of very dissimilar worlds.

The exploding universe discovered by Edwin Hubble and which until very recently seemed unimaginably complicated, this 'entire universe' totally beyond man's imagination, was found to be a negligible speck of matter in a much greater, much more complex flow of matter encompassing us.

In conclusion, we will briefly mention current concepts of the evolution of the universe at later stages than the first instants after the Big Bang and the synthesis of light elements at the beginning of the expansion.

After the first five minutes had elapsed, the temperature in the universe dropped to below one billion kelvin. By that time, all active processes involving elementary particles had ended and a long 'quiet' period had begun.

At this stage, the expanding plasma was still quite hot and opaque to radiation. The microwave background dictated the pressure in the plasma. This mixture of plasma and radiation contained density oscillations of small-amplitude: sound waves. Acoustic vibrations were the only process in the expanding plasma.

The theory of evolution of small perturbations in the expanding universe was constructed in 1946 by the Soviet physicist E. M. Lifshitz. He demonstrated that any deviations in the matter density from the uniform distribution in the high-temperature plasma at the early stage of expansion in the hot universe can exist only as sound waves. In this period, regardless of the choice of linear scale, the forces of gravitation cannot generate a growth in local plasma density, which would finally give rise to individual clouds or heavenly bodies. In other words, the mechanism of gravitational instability cannot work during this period. (The foundations of this theory had already been laid by Jeans at the beginning of the 20th century.)

Only after 3×10^5 years did the expanding plasma cool down to 4000 K and transform into neutral gas (via the recombination process). The neutral gas is practically transparent for the primordial background radiation. Now that the gas pressure is dictated only by the motion of neutral atoms, the elasticity of the gas drops sharply and the mechanism of the gravitational instability becomes effective. In 1964, one of the authors of this book (I. Novikov) showed how perturbations of sufficiently large wavelength in the hot plasma epoch can be driven by gravitation and evolve, after a sharp drop in pressure, into individual bodies. Further elaboration of the theory of gravitational instability was carried out by the schools of Ya. B. Zeldovich, E. M. Lifshitz, I. M. Khalatnikov, L. E. Gurevich and others in the USSR, and by P. Peebles, J. Bardeen and others in the USA. It has been understood in the last decade that weakly interacting particles, whose total mass may greatly exceed the mass of ordinary luminous matter, may have played an essential role in the evolution of the large-scale structure of the universe.

Individual galaxies and their clusters have formed, in all likelihood, in an epoch not very distant from ours, when all distances in the expanding universe were only several times smaller than today.

The problem of the formation of large-scale structure is closely linked with current observational cosmology and is being actively studied.

One of the most important problems of observational cosmology is the determination of the total mean density of matter in the universe. We have already mentioned that work is greatly hampered by the fact that the Universe contains forms of matter that are very difficult to observe: the 'dark', or 'hidden', mass.

Astronomers believed as recently as the 1950s that practically the entire matter of the universe is concentrated in luminous galaxies. The problem of calculating the mean density of matter could then be solved in the following manner. (1) Count the total number of galaxies in a sufficiently large volume of space. (2) Calculate the total mass of matter inside this volume by multiplying the mean mass of a galaxy by their total number. (3) Calculate the mean density by dividing the mass by the volume. A reliable determination of the mean matter density in galaxies was performed in this manner by the Dutch astronomer J. H. Oort in 1958. The estimate he reported was $\rho_1 = 2 \times 10^{-31}$ g/cm^3, assuming the Hubble constant to be 75 km/(s.Mpc).

This value for the density is 50 times less than the critical value separating the case of a permanently expanding infinite Universe from the case of a universe limited in space, in which expansion will be replaced in the future by contraction.

Later studies added nothing essential to Oort's estimate for matter in luminous galaxies. It was found, however, that extensive massive coronas of invisible matter exist around the visible galaxies outlined by the light-emitting stars. These coronas manifest themselves by their gravitation. This gravitation affects the motion of gas clouds far away from the boundaries of a luminous galaxy, and the motion of dwarf galaxies, which are satellites of large galaxies. The aggregate mass of invisible coronas is most probably much greater than that of the visible galaxies.

The 'hidden matter' is also revealed in studies of galaxy clusters. The total mass of a cluster, including the 'hidden mass' of matter between galaxies, is found by determining its gravitational potential. The potential can be measured by observing the velocities of motion of individual galaxies in the cluster and by measuring the temperature of the hot gas also present in the clusters. It was found that the 'hidden mass' is often greater than the total mass of the visible matter of a galaxy by more than an order of magnitude.

Taking the 'hidden mass' into account thus raises the total mean mass density of matter in the universe close to the critical value. We still do not know whether the total mean mass density is greater or less than this critical level.

The physical nature of the 'hidden mass' has not been clarified either. Partly, this could be the mass of faintly luminous stars which are practically invisible from a large distance. However, it seems more likely that the main part of the 'hidden mass' exists as a large number of elementary particles that interact very weakly with ordinary matter and thus manifest themselves only by gravitation. These could be neutrinos (if neutrinos have non-zero mass) or other particles similar to neutrinos that have remained in the universe since its expansion began.

We should never forget that the whole of extragalactic astronomy, which studies the greater universe using the most various methods, can be traced back to the work of the great Edwin Hubble, whose name and contribution are forever inseparable from this scientific achievement.

Chronology of main dates in the life
and work of Edwin Hubble

1889, Nov. 20 – Born in Marshfield, Mo., in the family of John Powell Hubble and Virginia Lee James

1897–1901 – First signs of interest in astronomy

1906 – Graduates from high school and enters Chicago University

1910 – Graduates from the university

1910–13 – Rhodes scholarship at Queen's College, Oxford University in Britain, where he studies law

1914 – Returns to Chicago University to prepare his thesis on astronomy

1916 – Publishes his first scientific paper

1917 – Defends his thesis and receives the PhD degree

1917–19 – Commands a batallion with the rank of major in the American Expeditionary Forces in France

1919 – Returns to the USA, leaves the Army and begins working at the Mount Wilson observatory

1922 – Publishes the principal papers on the study of diffuse nebulae in the Galaxy

1922–6 – Develops a classification of galaxies

1923 – Discovers the first Cepheid in the Andromeda nebula

1924 – Marries Grace Burke Leib

1926–9 – Publishes the results of studying the Andromeda and Triangulum nebulae. Formulates the concept of the island structure of the universe

1927 – Elected to the National Academy of Science of the USA

1928 – Elected to the Royal Astronomical Society of Great Britain

1929 – Discovers the law of red-shift (Hubble's Law)

1931–4 – Verifies the red-shift law (together with Humason)

1932 – Discovers globular clusters in the Andromeda nebula

1934 – Publishes galactic counts

1935 – Receives the Barnard medal of Columbia University

1935 – Elaborates, together with Tolman, methods of studying the nature of the red-shift

1936 – Publishes the book *The Realm of the Nebulae*

1937 – Publishes the book *The Observational Approach to Cosmology*

1938 – Receives the Bruce medal of the Astronomical Society of the Pacific

1939 – Receives the Franklin medal of the Franklin Institute

1940 – Receives the medal of the Royal Astronomical Society of Great Britain

1940 – First public speech calling for support for Britain in the war against Hitler's Germany

1942–5 – Works in the Ballistic Research Laboratory of the Aberdeen Proving Ground, Maryland

1946 – Receives the Medal for Merit for outstanding contribution to ballistics research during World War II

1947 – Elected to the Vienna Academy of Science

1949 – Elected to l'Institut Français (The French Academy of Science)

1949 – Obtains first photographic plates with the 200-inch reflector of the Mount Palomar Observatory

1953 – Publishes his last scientific paper, on high-luminosity variable stars in galaxies, known as Hubble–Sandage objects (jointly with Sandage)

1953, Sept. 28 – Dies suddenly, of a stroke, at the entrance to his house in San Marino, near Pasadena, California

Bibliography of Hubble's publications

Hubble's papers appeared not only as journal articles; typically, the Mount Wilson and Mount Palomar observatories reproduced them as reprints also mentioned in the list below – Communications, Contributions, Reprints. In some cases, Hubble's papers, especially his popularizing articles, were printed by several journals. Sometimes, papers of different size and similar content had identical titles. To save space, the bibliographic list mentions such titles once only but cites all references. Some sources were inaccessible to us, so we cite them without checking, copying Mayall's list accompanying his biography of Edwin Hubble.

1916

Twelve faint stars with sensible proper motions *Astronomical J.* **29**, 168.

The variable nebula NGC 2261 *Astrophysical J.* **44**, 190.

Changes in the form of the nebula NGC 2261 *Publications of the National Academy of Science* **2**, 230.

1917

Recent changes in the variable nebula NGC 2261 *Astrophysical J.* **45**, 351.

1920

The spectrum of NGC 1499 *Publications of the Astron. Soc. of the Pacific* **32**, 155.

The planetary nebula IC 2003 *Publications of the Astron. Soc. of the Pacific* **32**, 161.

Twelve new variable stars *Publications of the Astron. Soc. of the Pacific* **32**, 161.

Photographic investigations of faint nebulae *Publ. Yerkes Obs.* **4**, 69.

The color of the nebulous stars *Astrophysical J.* **52**, 8 (co-authored by F. H. Seares).

1921

Twelve new planetary nebulae *Publications of the Astron. Soc. of the Pacific* **33**, 174.

1922

A general study of diffuse galactic nebulae *Astrophysical J.* **56**, 162; *Mount Wilson Contr.* No 241.

The source of luminosity in galactic nebulae *Astrophysical J.* **56**, 400; *Mount Wilson Contr.* No 250.

Nova Z Centauri (1895) and NGC 5253 *Publications of the Astron. Soc. of the Pacific* **34**, 292 (co-authored by Lundmark).

1923

Density distribution in the photographic images of elliptical nebulae *Popular Astronomy* **31**, 644.

Messier 87 and Balanowsky's Nova *Publications of the Astron. Soc. of the Pacific* **35**, 261.

1925

NGC 6822, a remote stellar system *Astrophysical J.* **62**, 409; *Mount Wilson Contr.* No 304.

Cepheids in spiral nebulae *Popular Astronomy* **33**, 252; *Observatory* **48**, 139; *Science* **61**, 278.

1926

A spiral nebula as a stellar system: Messier 33 *Astrophysical J.* **63**, 236; *Mount Wilson Contr.* No 310.

Extragalactic nebulae *Astrophysical J.* **64**, 321; *Mount Wilson Contr.* No 324.

Non-galactic nebulae. I. Classification and apparent dimensions. II. Absolute dimensions and distribution in space (Abstract) *Publications of the Astron. Soc. of the Pacific* **38**, 258.

1927

The classification of spiral nebulae *Observatory* **50**, 276.

The nebulous envelope around Nova Aquilae N 3 *Astrophysical J.* **66**, 59; *Mount Wilson Contr.* No 335 (co-authored by J. C. Duncan).

Exploring the depths of space *News Serv. Bull. Carnegie Inst. Washington* **1**, No 6.

Density distribution in the photographic images of elliptical nebulae *Publications of the American Astron. Soc.* **5**, 63.

Cepheids in spiral nebulae *Publications of the American Astron. Soc.* **5**, 262.

1928

Novae or temporary stars *Leaflet Astron. Soc. Pacific*, No 14.

A spiral nebula as a stellar system: Messier 31 *Astrophysical J.* **69**, 103; *Mount Wilson Contr.* No 376.

1929

The structure of the universe – a clue *News Serv. Bull. Carnegie Inst. Washington* **6**, No 8, 49.

On the curvature of space *News Serv. Bull. Carnegie Inst. Washington* **6**, No 13, 49.

The exploration of space *Harper's Monthly Magazine* **158**, 732.

Preliminary estimate of the distance of the Coma cluster of nebulae *Publications of the Astron. Soc. of the Pacific* **41**, 247.

A relation between distance and radial velocity among extra-galactic nebulae *Proceedings of the National Academy of Science* **15**, 168; *Mount Wilson Comm.* No 105.

1930

Distribution of luminosity in elliptical nebulae *Astrophysical J.* **71**, 231; *Mount Wilson Contr.* No 398.

Velocity–distance relation among extra-galactic nebulae *Science* **72**, 407.

The nebulous envelope around Nova Aquilae 1918 *Popular Astronomy* **38**, 598 (co-authored by J. C. Duncan).

1931

The nebuous envelope around Nova Aquilae 1918 *Publictions of the American Astronomical Society* **6**, 365 (co-authored by J. C. Duncan).

The distribution of nebulae *Publications of the Astron. Soc. of the Pacific* **43**, 282.

The velocity–distance relation among extra-galactic nebulae *Astrophysical J.* **74**, 43; *Mount Wilson Contr.* No 427 (co-authored by M. L. Humason).

Nebulous objects in Messier 31 provisionally identified as globular clusters *Astrophysical J.* **76**, 44; *Mount Wilson Contr.* No 452.

1932

The surface brightness of threshold images *Astrophysical J.* **76**, 106; *Mount Wilson Contr.* No 453.

The distribution of extragalactic nebulae *Science* **75**, 24.

1934

The distribution of the extragalactic nebulae *Astrophysical J.* **79**, 8; *Mount Wilson Contr.* No 485.

Red-shifts in the spectra of nebulae, the Halley lecture delivered on 8 May 1934. Oxford: Clarendon Press, Oxford Univ. Press.

The award of the Bruce gold medal to Professor Alfred Fowler *Publications of the Astron. Soc. of the Pacific* **46**, 87.

The realm of the nebulae *Scientific Monthly* **39**, 139.

The velocity–distance relation for isolated extragalactic nebulae *Proceedings of the National Academy of Science* **20**, 264; *Mount Wilson Comm.* No 116 (co-authored by M. L. Humason).

1935

Two methods of investigating the nature of nebular red-shift *Astrophysical J.* **82**, 302; *Mount Wilson Contr.* No 527 (co-authored by R. C. Tolman).

Angular rotations of spiral nebulae *Astrophysical J.* **81**, 334; *Mount Wilson Contr.* No 514.

1936

Luminosity function of nebulae. I. The luminosity function of resolved nebulae as indicated by their brightest stars *Astrophysical J.* **84**, 158; *Mount Wilson Contr.* No 548.

The luminosity function of nebulae. II. The luminosity function as indicated by residuals in velocity–magnitude relation *Astrophysical J.* **84**, 270; *Mount Wilson Contr.* No 549.

Effects of red shifts on the distribution of nebulae *Astrophysical J.* **84**, 517; *Mount Wilson Contr.* No 557.

Effects of red shifts on the distribution of nebulae *Proceedings of the National Acedemy of Science* **22**, 621.

Ways of science *Bull. Occidental College, n. s.* **14**, No 1.

The Realm of the Nebulae. Oxford: Oxford Univ. Press.

A super-nova in the Virgo cluster *Publications of the Astron. Soc. of the Pacific* **48**, 108 (co-authored by G. Moore).

1937

The observational approach to cosmology (*Rhodes Memorial Lectures*). Oxford: Clarendon Press.

Red-shifts and the distribution of nebulae *Monthly Notices of the Royal Astron. Soc.* **97**, 506.

Our sample of the universe *Scientific Monthly* **45**, 481; Carnegie Inst. Washington, Supplementary Publ. No 33.

New comet *Harvard College Observatory. Announcement Card*, No 423.

1938

Adventures in cosmology *Leaflet Astron. Soc. Pacific*, No 115.

The nature of the nebulae *Publications of the Astron. Soc. of the Pacific* **50**, 97.

Das Reich der Nobel. Braunschweig: Verlag Friedr. Vieweg & Son.

Explorations in the realm of the nebulae *Cooperation in research*, Washington: Carnegie Inst. Washington, 91.

Jupiter X and Jupiter XI *Harvard College Observatory. Announcement Card*, No 465 (co-authored by S. Nicholson).

1939

Barred spirals *Publications of the American Astron. Soc.* **9**, 249.

The motion of the galactic system among the nebulae *Journal of the Franklin Inst.* **228**, 131.

The new stellar systems in Sculptor and Fornax *Publications of the Astron. Soc. of the Pacific* **51**, 40; *Mount Wilson Repr.* No 160 (co-authored by W. Baade).

Points of view: experiment and experience *Huntington Library Quarterly* **2**, No 3, 243.

The nature of the universe *Ann. Rep. Smithsonian Inst. for 1938*, 137.

1940

Problems of nebular research *Scientific Monthly* **51**, 141.

1941

Supernovae *Publications of the Astron. Soc. of the Pacific* **53**, 141.

The direction of rotation of spiral nebulae *Science* **93**, 434.

The role of science in a liberal education *University and the Future of America*. Stanford: Stanford University Press, p. 137.

Zwicky's system in Sextans and Leo *Scientific Monthly* **52**, 486.

The problem of the expanding universe *Science* **95**, 212; *American Scientist* **30**, 99.

1943

The direction of rotation of spiral nebulae *Astrophysical J.* **97**, 112; *Mount Wilson Contr.* No 674.

The problem of the expanding universe *Ann. Rep. Smithsonian Inst. for 1942*, 119; *Publ. Smithsonian Inst.* No 3707; *Sci. Monthly* **56**, 15.

1945

The problem of the expanding universe *Science in Progress. Third Series.* New Haven: Yale Univ. Press, p. 22

The Exploration of Space: The Scientists Speak, ed. W. Weaver. New York: Boni & Baer, p. 37

1946

The exploration of space *Popular Astronomy* **54**, 183.

1947

The 200-inch telescope and some problems it may solve *Publications of the Astron. Soc. of the Pacific* **59**, 153.

1948

Mars and the 200-inch telescope *Monthly Notes of the Astronomical Society of Southern Africa* **7**, 15.

The greatest of all telescopes *Listener* **15**, 1025.

1949

First photographs with the 200-inch Hale telescope *Publications of the Astron. Soc. of the Pacific* **61**, 121; *Mount Wilson and Palomar Repr.* No 4.

Five historic photographs from Palomar *Scientific American* **181**, 32.

1950

Fotografías históricas del Cielo tomadas en el observatorio Palomar *Boletin de Ciencia y Technologia*, No 2, 22.

The 200-inch Hale telescope and some problems it may solve *Ann. Rep. Smithsonian Inst. for 1949*, 175; *Smithsonian Inst. Publ.* No 3999.

1951

Explorations in space: the cosmological program for the Palomar telescopes *Proc. Amer. Philosoph. Soc.* **95**, 461; *Mount Wilson and Palomar Repr.* No 55.

1953

The brightest stars in extragalactic nebulae I. M 31 and M 33 *Astrophysical J.* **118**, 353; *Mount Wilson and Palomar Repr.* No 107 (co-authored by A. Sandage).

1954

The law of red-shifts *Monthly Notices of the Royal Astronomical Soc.* **113**, 658; *Mount Wilson and Palomar Repr.* No 131.

The Nature of Science and Other Lectures. San Marino, Calif.: The Huntington Library.

Fünf historische Himmelphotographien vom Mt. Palomar *Naturwiss. Rundschau* **7**, 137.

1958

The Realm of the Nebulae New Haven and London: Yale Univ. Press.

1982

The Realm of the Nebulae New Haven and London: Yale Univ. Press.

References

The list that follows includes the main publications which, together with the archived documents, the memoirs of Mrs Helen Lane, Edwin Hubble's sister, the interview of Dr Sandage and some other material, were the sources for writing this biography of E. Hubble. The bibliography is appended with a list of books recommended to those who would like to read up on the history of astronomy in the 20th century and on cosmological problems *per se*.

Adams, W.S. (1923). *Publications of the Astron. Soc. of the Pacific* **35**, 290.
Adams, W.S. (1954). *Publications of the Astron. Soc. of the Pacific* **66**, 267.
Adams, W.S. (1954). *Observatory* **74**, 32.
Aitken, R.G. (1939). *Publications of the Astron. Soc. of the Pacific* **51**, 5.
American Armies and Battlefields in Europe. A History, Guide and Reference Book (1938). Washington: US Government Print. Office.
Anderson, J.A. (1939). *Publications of the Astron. Soc. of the Pacific* **51**, 24.
A Source Book in Astronomy and Astrophysics, 1900–1975 (1979). (eds. K.K. Lang and O. Gingerich) Cambridge, Mass. and London: Harvard Univ. Press.
Baade, W. (1948). *Publications of the Astron. Soc. of the Pacific* **60**, 230.
Babcock, H. D. (1938). *Publications of the Astron. Soc. of the Pacific* **50**, 87.
Berendzen, R. & Hoskin, M. (1971). *Leaflet Astron. Soc. Pacific* No. 504.
Bowen, I. (1954). *Science* **119**, 204.
Bowen, I. (1972). *Quarterly J. Royal Astron. Soc.* **14**, 235.
Canby, T.Y. (1974). *National Geographic* **145**, 626.
Chant, C.A. (1953). *J. Royal Astron. Soc. Canada* **47**, 225.
De Sitter, W. (1917). *Monthly Notices of the Royal Astron. Soc.* **78**, 3.
De Sitter, W. (1930). *Bull. Astron. Inst. Netherlands* **5**, 157.
De Cicco, D. (1986). *Sky and Telescope* **71**, 347.
Dose, A. (1926). *Astronomische Nachrichten* **229**, 157.
Douglas, A.V. (1957). *The Life of Arthur Stanley Eddington.* London: Thomas Nelson & Sons.
Friedmann, A. (1922). *Zeitschrift für Physik* **10**, 377.
Gingerich, O. (1987). *J. Royal Astron. Soc. Canada* **81**, 113.
Hall, J.S. (1970). *Sky & Telescope* **39**, 84.
Harbord, J.G. (1936). *The American Army in France. 1917–1919.* Boston: Little Brown.
Hart, R. & Berendzen, R. (1971). *Journal for History of Astronomy* **2**, 109.
Hetherington, N.S. (1972). *Quarterly J. Royal Astron. Soc.* **13**, 25.
Hetherington, N.S. (1983). *Nature* **306**, 727.
Hetherington, N.S. (1986). *Nature* **319**, 189.
History of American Field Service in France. Friends of France. 1914–1917, 1920 (1920). Vol. **3** Boston & New York: Houghton Miflin.
Hoskin, M. (1982). *Stellar Astronomy. Historical Studies.* London: Science History. London: Chalfont St. Giles.
Humason, M.L. (1929). *Proc. Nat. Acad. Science* **15**, 167.
Humason, M.L. (1931). *Astrophysical J.* **74**, 35.

Humason, M.L. (1934). *Publications of the Astron. Soc. of the Pacific* **46**, 290.

Humason, M.L. (1935). *Publications of the Astron. Soc. of the Pacific* **47**, 223.

Humason, M.L. (1954). *Monthly Notices of the Royal Astron. Soc.* **114**, 291.

Huxley, A.L. (1969). *Letters of Aldous Huxley.* (ed. G. Smith) London: Chatto & Windus.

Jaffe, B. (1958). *Men of Science in America,* revised edn. New York: Simon & Schuster.

Jeffers, H.M. (1937). *Publications of the Astron. Soc. of the Pacific* **49**, 271.

Learner, R. (1986). *Sky & Telescope* **71**, 349.

Lemaitre, G. (1929). *Annales de la société scientifique de Bruxelles* **47**, Serie A, 49; *Monthly Notices of the Royal Astron. Soc.* **41** 483.

Lundmark, K. (1924). *Observatory* **47**, 279.

Lundmark, K. (1924). *Monthly Notices of the Royal Astron. Soc.* **84**, 747.

Lundmark, K. (1925). *Monthly Notices of the Royal Astron. Soc.* **85**, 865.

Lundmark, K. (1927). *Meddelanden Astron. Obs. Upsala,* No. 30.

Mayall, N.U. (1954). *Sky & Telescope* **13**, 78.

Mayall, N.U. (1970). *Biographical Memoirs of the National Academy of Science in the United States of America* **41**, 175.

Order of Battle of the United Land Forces in the World War: American Expeditionary Forces. (1931). Washington: U.S. Government Print. Office.

Osterbrock, D.E. (1976). *Sky & Telescope* **51**, 91.

Paddock, G.F. (1916). *Publications of the Astron. Soc. of the Pacific* **28**, 109.

Parker, B. (1986). *Sky & Telescope* **72**, 227.

Payne-Gaposchkin, C. (1984). *An Autobiography and other Recollections.* Cambridge: Cambridge Univ. Press.

Plummer, H.C. (1940). *Monthly Notices of the Royal Astron. Soc.* **100**, 342.

Reynolds, J.H. (1927). *Observatory* **50**, 185.

Richardson, R.S. (1948). *Publications of the Astron. Soc. of the Pacific* **60**, 215.

Robertson, H.P. (1928). *Philosophical Magazine* **5**, Seventh series, 835.

Robertson, H.P. (1954). *Publications of the Astron. Soc. of the Pacific* **66**, 120.

Russell, H.N. (1925). *Scientific American* **132**, 165.

Russell, H.N. (1929). *Scientific American* **140**, 504.

Russell, H.N. (1932). *Scientific American* **146**, 14.

Sandage, A.R. (1954). *Astronomical Journal* **59**, 180.

Sandage, A.R. (1961). *The Hubble Atlas of Galaxies.* Washington: Carnegie Inst. Publ. No. 618.

Schlesinger, F. (1935). *Publications of the Astron. Soc. of the Pacific* **47**, 175.

Shapley, H. (1929). *Publications of the National Academy of Science* **15**, 565.

Seeley, D. & Berendzen, R. (1978). *Mercury* **7**, 67.

Slipher, V.M. (1915). *Popular Astronomy* **23**, 21.

Slipher, V.M. (1916). *Bull. Lowell Obs.* **2**, No. 58, 56.

Slipher, V.M. (1917). *Proc. Amer. Philosoph. Soc.* **56**, 406.

Smith, R.W. (1982). *The Expanding Universe. Astronomy's 'Great Debate', 1900–1931.* Cambridge: Cambridge Univ. Press.

Strömberg, G. (1925). *Astrophysical J.* **61**, 353.

Truman, O.H. (1916). *Popular Astronomy* **24**, 111.

Weart, S.R. & De Vorkin, D.H. (1982). *Sky & Telescope* **63**, 124.

Weyl, H. (1923). *Phys. Zeitschrift* **24**, 230.

Wirtz, C. (1917). *Astronomische Nachrichten* **206**, 109.

Wirtz, C. (1922). *Astronomische Nachrichten* **215**, 349.

Wirtz, C. (1922). *Astronomische Nachrichten* **216**, 451.

Wirtz, C. (1924). *Astronomische Nachrichten* **222**, 22.

Wright, W.H., Einarsson, S. & Jeffers, H.M. (1937). *Publications of the Astron. Soc. of the Pacific* **49**, 270.

Young, R.K. & Harper, W. E. (1916). *J. Royal Astron. Soc. Canada* **10**, 134.

Zwicky, F. (1929). *Publications of the National Academy of Science* **15**, 773.

*   *   *

Baade, W. (1963) *Evolution of Stars and Galaxies* (ed. C. Payne-Gaposchkin) Cambridge, Mass.: Harvard Univ. Press.

Barrow, J. & Silk, J. (1983) *The Left Hand of Creation.* New York: Basic Books.

Einstein, A. (1967). *Collected Works.* Vols 2 & 4 Moscow: Nauka (in Russian).

Frenkel, V.Ya. (1974). *New materials concerning the Einstein–Friedmann debate on relativistic cosmology.* Einstein Collected Papers, 1973 Moscow: Nauka.

Friedmann, A.A. (1966). *Collected Works.* Moscow: Nauka.

Novikov, I.D. (1983). *Evolution of the Universe.* Cambridge: Cambridge University Press.

Novikov, I.D. (1988). *How the Universe Exploded.* Moscow: Nauka (in Russian).

Novikov, I.D. (1990). *Black Holes and the Universe.* Cambridge: Cambridge University Press.

Penzias, A.A. (1979) The origin of the elements. Nobel lecture. 8 December 1978. *Rev. Mod. Phys.* **51**, 430.

Silk, J. (1980). *The Big Bang. The Creation and Evolution of the Universe.* San Francisco: W. H. Freeman.

Struve, O. & Zebergs, V. (1962). *Astronomy of the 20th Century.* New York: Macmillan.

Whitney, C.A. (1971). *The Discovery of our Galaxy.* New York: Alfred Knopf.

Wilson, R.W. (1979) The cosmic microwave background radiation. Nobel lecture. 8 December 1978. *Rev. Mod. Phys.* **51**, 440.

Yefremov, Yu.N. (1984). *To the Depths of the Universe.* Moscow: Nauka (in Russian); (1990). *In die Tiefen des Weltalls* Leipzig: Mir–Teubner.

Index of names

Abbot Ch. G. – 15
Adams W. S. – 24, 31, 51, 105
Al-Sufi – 27
Albrecht A. – 157
Alexander S. – 25
Alfvén H. – 78, 108
Alpher R. – 143, 144, 149
Ames A. – 85
Apian P. – 85
Arliss G. – 82
Arp H. C. – 80
Arrhénius S. A. – 76

Baade W. – 26, 41, 43, 45, 46, 65, 80, 81, 86, 87, 98, 103, 128
Babcock H. D. – 79
Bach J. S. – xv
Balanovsky I. A. – 104, 114
Bardeen J. – 158, 168
Barnard E. E. – 15, 32, 78
Barrow J. D. – 164
Bateson W. – 76
Baum W. A. – 115
Berendsen R. – 25
Bernoulli – xv
Bethe H. – 78
Bigourdan G. – 21
Bohr N. – 6
Bolin, Karl – 28
Bowen I. S. – 105
Bruce C. W. – 78
Burke, G. (Mrs Edwin Hubble) – xi, 33
Burke B. – 147
Bush V. – 106

Campbell W. S. – 8, 21, 25, 77, 107
Carnegie A. – 15
Carter B. – 164
Cassegrain N. – 68
Cassini G. D. – 27
Chamberlin Th. C. – 15

Chandrasekhar S. – 78
Chibisov G. S. – 158
Chretien H. – H. 46
Christie W. – 66, 98
Cleaveland – 38
Compton A. – 93
Copernicus N. – viii, 76, 85, 121
Curtis H. D. – 21, 27, 28, 35, 42

Darwin, C. – 83, 117
Darwin, G. – 76, 117
De Sitter, W. – 46, 55, 57, 57, 67, 82
De Vaucouleurs G. – 131, 132
Dicke R. H. – 147, 164
Dirac P. – 164
Disney W. – 84
Dobrynin A. – xii
Doroshkevich A. G. – 149
Dose A. – 56
Dubridge A. – 111
Duke of Edinburgh – 120
Duncan J. C. – 16, 18, 28, 33, 37, 44, 105
Dunham T. – 98
Dymnikova I. G. – 157
Dyson J. – 54

Eddington, Sir Arthur – 36, 43, 54, 60, 65, 70
Efimov Yu. S. – xii
Efremov Yu. N. – 132
Ehrenfest P. – 59, 164
Einstein A. – 52, 54, 55, 59, 75, 89, 133, 156

Fairey, Sir Charles Richard – 84, 120
Fermi E. – 101
Fild J. – 150
Fisher R. – 127, 130
Flemsteed J. – 27
Fock V. A. – 55
Fowler A. – 76

Fowler W. – 78
Franco F. – 89
Franz J. – 28
Franz Ferdinand, Archduke – 11
Friedmann A. A. – 58, 65, 98, 142
Fritzsche H. – 120
Frost E. – 9, 35, 51

Galileo Galilei – viii, 76, 85, 121
Gamow G. – 144, 148, 164
Gerasimovich B. P. – 114
Gingerich O. – 26
Ginzburg V. L. – 151
Gliner E. B. – 157
Griffin R. – 122
Gurevich L. E. – 157, 168
Gutenberg J. – 84
Guth A. – 157, 158

Hadamard J. S. – 76
Hale G. E. – 12, 15, 22, 25, 77, 84
Halley E. – 76
Harper W. E. – 11, 50
Hart R. – 25
Hawking S. – 151, 158, 164
Hermann R. – 144
Herschel W. – 7, 26, 27, 32, 90, 121
Hertzsprung E. – 20, 30, 79
Hetherington N. – 40
Hevelius J. – 85
Hewish A. – 78
Hindenburg, von P. – 89
Hitchcock J. – 150
Hitler, A. – 84, 89, 94
Hodge P. W. – xii, 133
Hofflighter D. – 132
Holden E. S. – 79
Holmberg E. – 98
Hooker J. D. – 15, 16
Hoskin M. – 40
Hoyle F. – 115
Hubble, Richard – 1
Hubble, Grace – 83
Huchra J. P. – 139
Huggins, Sir William – 16, 48
Humason M. – 16, 18, 28, 45, 46, 61, 64, 66, 68, 75, 88, 105, 115
Hunt M. – 16

Huntington H. – 84
Huxley A. – xv, 5, 83, 84, 113
Huxley H. – 82
Huxley J. – 82, 113, 114

Jackson – 121
Jeans J. H. – 21, 25, 40, 43, 44, 168
Jeffers, Dr – 78
Jenkins L. – 9
Jordan F. C. – 10

Kaidanovsky N. L. – 150
Kant I. – 164
Kapteyn J. – 51
Kardashov N. S. – 138, 142
Keeler J. – 33
Kent R. – 95
Kepler, J. – 49, 85
Kerr P. H. – 84
Khaikin S. E. – 150
Khalatnikov I. M. – 151, 168
Kharadze E. K. – xii
Khlopov M. Yu. – 155
Kholopov P. N. – xii
King Arthur – 45
Kippenhahn R. – 77
Kirzhnits D. A. – 157
Klumpke-Roberts D. – 22
Krutkov Yu. A. – 59
Kulikovsky P. G. – xii

Lane, H. – xii, 4, 6, 34
Lang K. K. – 26
Lattre de Tassigny J. – 101
Lavrova N. B. – xii
Leavitt, H. – 29
Lemaitre J. – 60, 76
Lermontov, M. – ix
Lifshitz E. M. – 167, 168
Lindblad B. – 85, 88
Linde A. D. – 157, 158, 166
Lloyd George D. – 84
Longair M. – xii
Lorentz H. A. – 75
Lowell P. – 47
Lukash V. N. – 158
Lundmark K. – 21, 24, 25, 28, 52, 54, 56

Lysenko T. D. – 114

Mach E. – 164
Maeterlinck M. – 4
Malthus T.M. – 7
Maltsev V. – 38, 63
Marius S. – 27
Markov A. M. – 151
Matthews T. – 134
Mayall N. U. – xi, 17, 82, 85, 89, 121
McCrea W. H. – 76
McKellar A. – 150
McLaughlin D. B. – 80
Melanchthon F. – 85
Merrill P. W. – 107
Messier Ch. – 27, 32, 35
Michelson A. – 6
Miller J. – 49
Millikan R. A. – 6, 75
Milne E. A. – 63, 76, 115
Mineur X. – 128
Minkowski R. – 105
Moor G. – 105
Morgan T. H. – 76
Moulton F. R. – 6, 15, 48
Mukhanov V. M. – 158

Nernst W. – 76
Newton I. – 7, 85
Nicholson S. – 40, 41

Oort J. H. – 76, 85, 169
Oppenheimer J. R. – 101, 107
Osterbrock D. – 17
Ostriker J. – 151

Paddock J. – 51, 53
Pariisky Yu. N. – 140
Payne-Gaposchkin C. – 31
Pease F G – 9, 16, 33, 64
Peebles P. – 147, 151, 168
Penzias A. – 78, 146, 149
Perrine Ch. D. – 32,
Philips T. E. R. – 43
Pickering E. – 9
Plackett H. H. – 9
Plankett E. J. M. – 84

Pliny Sr – 85
Plummer H. G. – 90,
Pollock E. T. – 36
Purbach G. – 85

Rastorguev A. S. – xiv
Rees M. J. – 151, 164
Regiomontanus J. – 85
Reynolds J. H. – 21, 26,
Rheil D. – 97
Rhodes, C. – 6
Riccioli G. B. – 85
Ritchey G. – 28, 29, 33
Roberts I. – 10, 22, 27, 32, 48, 90
Robertson H. – 60
Roll P. – 147
Roosevelt F. D. – 91, 94, 95
Ross, Lord – 25, 32, 90
Rozental I. L. – 162, 164
Russell N. H. – 36, 38, 57, 64, 108
Rutherford, Sir Ernest – 6, 70, 76
Ryle M. – 77

Sacrobosco J. – 85
Sagan C. – 108
Samus N. N. – xii
Sandage A. – xii, 8, 33, 81, 91, 116, 122, 129, 131, 134
Sanford R. F. – 16, 28
Shain G. A. – vii
Scheiner Ch. – 28,
Schiaparelli G. – 47
Schlesinger F. – 36
Schliemann H. – 47
Schmidt M. – 134
Schoenfeld – 90
Seares S. H. – 30, 41,
Shain G. A. – ix
Shapley H. – 24, 26, 29, 30, 31, 35, 37, 40, 46, 57, 74, 76, 79, 82, 86, 87, 108
Shklovsky I. S. – 44, 104, 134, 145, 150
Shmaonov T. A. – 150
Silk J. – 164
Silliman H. E. – 76
Simon, Colonel – 95

Slipher V. M. – 9, 20, 21, 22, 35, 45, 47, 51, 60, 82
Smart W. M. – 43
Smith H. – 135
Sommerfeld A. – 74
Spaatz C. – 101
St. John, C. – 45
Starobinsky A. A. – 157, 158, 166
Stebbins J. – 36, 38
Steinhardt P. – 157, 158
Stratton F. J. M. – 44
Stravinsky I. – 83
Struve, generations of – xv
Struve O. – 82
Strömberg G. B. – 56, 98
Sunyaev R. A. – 130

Tammann G. A. – 131, 134
Tedder A. W. – 101
Thackeray A. D. – 129
Thaddeus P. – 150
The Seven Samurai – 131
Thompson, Lord – 76
Thorne K. S. – xiv, 8
Tolman R. – 63, 72, 75, 110
Truman O. H. – 11, 50
Tuchman B. – 11
Tully R. – 130
Turner M. S. – 158

Umbaraeva N. D. – 138

Urey H. – 101

Van den Bergh S. – 129, 132
Van Maanen A. – 9, 39, 40, 46
Vavilov S. I. – xiii, 47
Vilenkin A. – 158

Walpole, Sir Hugh Seymour – 84
Wasson L. – 6
Weinberg S. – 147, 151
Weyl H. – 60
Wheeler J. A. – 164, 166
Whitford A. E. – 116
Wilkinson D. – 147
Wilson O. – 98
Wilson R – 78, 98, 146
Wilson W. – 11
Wirtz C. – 52, 53, 55
Wolf J. A. – 21, 150
Wolf M. – 21, 32, 33, 107
Wright F. – 33

Yerkes Ch. T. – 9, 15, 86
Young R. K. – 11, 50

Zeldovich Ya. B. – 128, 130, 145, 151, 155, 156, 164, 168
Zelmanov A. L. – 151
Zhdanov A. A. – 115
Zhukov G. – 101
Zwicky F. – 67, 81 , 105

Index of celestial objects

3C 48 quasar – 131
3C 273 quasar – 132

Andromeda galaxy (nebula, M31) –
21, 27, 28, 30, 36, 37, 40, 42, 43, 45,
48, 49, 61, 80, 99, 111, 119, 128
asteroid No. 2069 – 122

Coal Sack – 19
Crab nebula – 44, 45

Fornax system – 87

Galaxy – viii, 19, 27, 42, 46, 108, 119,
127, 138, 142
Gould's belt – 19
Great Attractor – 131

Hubble–Sandage objects – 117
Hyades – 126
Hydra cluster – 118, 131

IC 2003 – 18

Large Magellanic Cloud – 104, 128
Local Group of galaxies – 47

M32 – 21, 99
M33 (Triangulum nebula) – 33, 36,
37, 39, 41, 42, 43
M51 – 39, 41, 103
M74 – 42
M81 – 37, 39, 41, 103, 116, 117
M87 – 104, 113
M101 – 31, 37, 41, 103, 116, 117
Magellanic Clouds – 24, 32, 54, 79
Mare Marginis – 122
Milky Way – viii, 15, 17, 19, 20, 49,
71, 87, 121

NGC 147 – 99
NGC 1499 – 18
NGC 185 – 99
NGC 205 – 46, 99

NGC 2261 – 112
NGC 2403 – 37, 116, 117
NGC 3177 – 103
NGC 3184 – 105
NGC 3359 – 113
NGC 3977 – 103
NGC 4038 – 105
NGC 4151 – 18
NGC 4273 – 105
NGC 4594 – 49
NGC 4632 – 103
NGC 4699 – 103
NGC 5204 – 113
NGC 5253 – 104
NGC 6822 – 32, 33, 36, 87
NGC 7619 – 61, 64
Neptune – 66

Pegasus cluster of galaxies – 61, 66

R Monocerotis – 9, 112
Regulus – 66
RR Lyrae – 111

S Andromedae – 27
Saturn – 111
Sculptor system – 87
Seyfert galaxies – 18
Small Magellanic Cloud – 29, 35, 128
Solar System – 48, 53, 108, 115
Sun – viii, 15, 50, 56, 61
supernova of 1885 – 28, 43
supernova of 1987 in Large
Magellanic Cloud – 139

Ursa Major cluster of galaxies – 66,
68, 116

Variable A in M 33 – 117
variable B in M 33 – 117
Virgo cluster of galaxies – 49, 62, 64,
66, 103, 116, 129, 132

Z Centauri – 24, 104

Subject index

10-inch astrograph – 17, 18
10-inch refractor – 86
40-inch reflector – 86
60-inch reflector – 16, 18, 19, 29, 33, 86, 98
100-inch reflector – 15–17, 20, 32, 33, 68, 71, 86, 99, 132
200-inch reflector – 45, 73, 77, 79, 107, 110

Aberdeen Proving Ground – 95
age of the Earth – 144
age of the universe (*see also* lifetime of the universe) – 144
American Astronomical Society – 38, 63, 86
American Institute of Physics – 109
anthropic principle – 163
apparent magnitude *vs* red-shift dependence – 72, 73, 136
Association for the Advancement of Science – 39
Astro Space Centre of the Lebedev Physics Institute – 141
Astronomical Society of the Pacific – 78
average density of matter – 138

Baade window – 87
Barnard Gold Medal – 78
Bell Laboratories – 146
Big Bang – 69, 143, 144, 148, 148, 167
binding energy – 160
bright and dark nebulae – 17, 19
British Royal Astronomical Society – 118, 121

Calibration of distance indicators – 126
California Institute of Technology – 67, 75, 77, 85, 107, 111

Carnegie fund – 15
Carnegie Institution – 15, 34, 111
Cassegrain hole – 86
Cepheids – 29, 30, 31, 35, 37, 40, 42, 66, 68, 80, 87, 111, 116, 119, 125, 132, 137, 152, 167
clusters of galaxies – 63, 66, 87, 111, 127, 130, 154
cold birth of the Universe – 143, 145
compactified dimensions – 166
convective layers of stars – 161
Cordoba Observatory – 79
Cordova – 32
Cormac lecture – 120
Cosmic Background Explorer (COBE) – 153
cosmological equation – 52
Cosmological models – 57
Coulomb's law – 164
critical density problem – 154
Crossley 36-inch reflector – 33

Dark nebulae – 18, 19
dark matter – 168
dark nebulosity – 20
Darwin lecture – 117
density – 158
De Sitter's empty Universe – 59
deuterium – 145, 160
deviations from uniformity – 154
diffuse nebulae – 20, 87
discovery of the hot Universe – 143
distance indicators – 126
Doppler shift – 130

ECHO satellite – 146
Einstein's model – 58
Einstein's theory – 54
Einstein–de Sitter model – 132
elliptical galaxies – 22, 25, 113
elliptical objects – 45
elongated nebulae – 20

evolution of small perturbations – 167
exploding eternity – 167
extragalactic nebulae – 19, 20, 21, 23

First moments of the Big Bang – 155
formation of the large-scale structure – 168
Franklin Institute – 81
French Institute – 120
Friedmann's equations – 57, 149
Friedmann's model – 58, 154
fundamental properties of the Universe – 152

Galactic nebulae – 19, 25
Gedanken experiment – 159
general relativity – 51, 57
globular clusters – 18, 46, 81, 99, 111, 119, 127
globular nebulae – 21
Goddard High Resolution Spectroscope – 139
Gould's belt – 19
Grand Unification – 154
gravitation constant – 155
Greenwich Observatory – 78

Hale telescope – vii
Harvard Observatory – 79
Heidelberg – 32
helium abundance – 144
Hertzsprung–Russell diagram – 127
hidden matter – 168
HII regions – 100, 129
horizon problem – 153, 157
horn reflector – 146, 149
'hot' scenario – 144
Hot Universe – 139
Hubble constant – 63, 67, 69, 111, 129, 132, 144
Hubble expansion – 131
Hubble family – 1
Hubble's Law – 63, 133
Hubble Space Telescope – 136
Huntington library – ix
hydrogen atom – 159

Inflation – 156
International Astronomical Union (I.A.U) – 21, 24, 45, 82
International Ultraviolet Explorer Satellite – 139
ionised hydrogen clouds – 127
irregular nebulae – 21
irregulars – 22

KRT-10 – 141
K-term – 51, 53, 56, 61, 62

Largest red-shifts – 136
Λ-term – 58, 60, 69
Lick Observatory – 21, 33, 51, 79, 85
lifetime of a typical star – 163
lifetime of the Universe (*see also* age of the Universe) – 163
'long' scale of extragalactic distances – 132
Lowell observatory – 49

MacCormic observatory – 97
mean matter density – 169
megaparsec (Mpc) – 62, 129,
Messier catalogue – 35
method of photometric distance indicators – 140
method of trigonometric parallax – 140
microwave background radiation (*see also* relic electromagnetic radiation) – 58, 130, 146, 151
Molton–Chamberlin model – 48
monopole problem – 155
Morrison public lecture – 108
Mount Palomar – viii, 76, 86, 116
Mount Wilson Observatory – viii, 15, 17, 25, 33, 34, 43, 56, 61, 75, 83, 118

Naval Observatory in Washington – 36
nebulae, galactic – 20
nebulae, irregular – 23
nebulous stars – 17
negative pressure – 155
National Academy of Sciences of the USA – viii, 36, 106

Newton telescope – vii
Newton's law – 85, 156, 164
Nobel Prize for physics – 147
novae – 37, 80, 116, 127

O and B classes – 43
one-metre refractor – 15
open clusters – 99
origin of small primary fluctuations
of matter density – 158

P. K. Sternberg Astronomical
Institute – 135
Paris Observatory – 79
parsec (pc) – 9
Pearl Harbor – 98
period–luminosity curve for Cepheids
– 30, 43
photoelectron multipliers – 119
Planck density – 155, 166
Planck energy – 158
Planck's constant – 155
Planck's formula – 147
planetary nebula 'Saturn' in
Aquarius – 86
planetary nebulae – 18, 19
population I, stellar – 99
population II, stellar – 99
prediction of the antiproton – 148
'primer' of the expansion – 156
primeval matter – 145
primordial (relic) electromagnetic
radiation – 145

Quantum fluctuations of the vacuum
– 155, 166
quasars – 134

Radio-telescopes, cosmic – 141
RADIOASTRON – 141, 142
radio-interferometers – 141
red-shift – 67, 98
red-shift law – 115, 116, 119
relativistic effects – 136
relic electromagnetic radiation (*see
also* microwave background
radiation) – 146, 151
RELIKT satellite – 153

Reynolds' classification – 21
Ritchey–Chretien telescope – 47
Rockefeller Foundation – 111
Royal Astronomical Society in
London – 41, 43, 48, 53

Saint Petersburg – 47
Salyut 6 – 141
scale of extragalactic distances –
126
Schmidt camera, 48-inch – 86, 105,
113
Schmidt telescope, 18-inch – 105
secondary indicators – 127
Seyfert galaxies – 18
Shain reflector telescope – vii
Shapley–Ames catalogue – 85
'short' scale of extragalactic distances
– 132
singularity – 158
solar mass – 88
Space Research Institute – 153
Space Telescope Research Institute –
137
spectrum of the microwave
background radiation – 147
spherical aberration – 139
spiral nebulae – 20, 53
spiral objects – 25, 45
spirals, barred – 22
spirals, normal – 22
star clusters – 87
state of quantum foam – 166
supergiant stars – 87, 127
supernovae in Virgo – 28, 105, 127

Theory of gravitational instability –
168
triangulation of the Universe –
142
Troy – 47
Tully–Fisher method – 130

Uniform Hubble flow – 131

Vacuum foam – 167
vacuum-like states – 155
variable nebula – 18

variables – 29, 37, 100, 116
variable stars in the Andromeda
nebula – 116
variable stars in the Triangulum
nebula – 115
variation of the gravitational
constant, imaginary – 161

Victoria Observatory – 50
virtual pairs – 155

Yerkes Observatory – 15, 18, 79

Zone of avoidance – 70